Between Nucleus and
Cytoplasm

Between Nucleus and Cytoplasm

Paul S. Agutter

Reader in Cell Biology, Department of Biological Sciences, Napier College, Edinburgh

CHAPMAN AND HALL

London • New York • Tokyo • Melbourne • Madras

UK	Chapman and Hall, 2–6 Boundary Row, London SE1 8HN
USA	Chapman and Hall, 29 West 35th Street, New York NY10001
JAPAN	Chapman and Hall Japan, Thomson Publishing Japan, Hirakawacho Nemoto Building, 7F, 1-7-11 Hirakawa-cho, Chiyoda-ku, Tokyo 102
AUSTRALIA	Chapman and Hall Australia, Thomas Nelson Australia, 480 La Trobe Street, PO Box 4725, Melbourne 3000
INDIA	Chapman and Hall India, R. Seshadri, 32 Second Main Road, CIT East, Madras 600 035

First edition 1991

© 1991 P. S. Agutter

Typeset in Plantin 10/12pt by Photoprint, Torquay
Printed in Great Britain at the University Press, Cambridge

ISBN 0 412 32180 7 (HB)
 0 412 32190 4 (PB)

British Library Cataloguing in Publication Data

Agutter, Paul S.
 Between nucleus and cytoplasm.
 1. Organisms. Cells
 I. Title
 574.87

 ISBN 0–412–32180–7
 ISBN 0–412–32190–4 pbk

Library of Congress Cataloging-in-Publication Data

Agutter, Paul S., 1946–
 Between nucleus and cytoplasm/Paul S. Agutter: diagrams by Jen Harvey.
  p. cm.
 Includes bibliographical references.
 ISBN 0–412–32180–7. — ISBN 0–412–32190–4 (pbk.)
 1. Cell nuclei. 2. Cytoplasm. 3. Biological transport. 4. Cellular control mechanisms. I. Title.
 QH595.A36 1990
 574.87′5—dc20   89–70876
                                                    CIP

Contents

Preface

In 1980, two pioneers of the study of nucleocytoplasmic transport, Philip
Paine and Samuel Horowitz, wrote that our knowledge of the subject was
'anecdotal' and lacked any specific concepts or principles of its own. Now, a
decade later, this is no longer true. Nucleocytoplasmic transport has
developed into one of the most active and exciting areas of research in cell
biology, and its significance for cell biology as a whole has become clear to
everyone in the field. It throws important light on the control of major
cellular functions such as protein biosynthesis, and on the mechanisms
involved in cellular differentiation. Moreover, it is helping to bring about an
invaluable synthesis of ultrastructural, biochemical, molecular biological and
biophysical approaches to the study of cellular organization and function.

This dramatic change has been brought about largely by technical
advances: improved methods of gene manipulation and sequencing, of
monoclonal antibody production and use, of protein chemistry, of microscopy and of cell manipulation including microinjection methods. It has also
received important stimuli from apparently unrelated developments in
molecular biology, such as the search for sequence-specific DNA binding
proteins and the identification of their binding domains. Recently, it has
become increasingly integrated with some well-established areas of biochemistry, such as the study of nuclear envelope structure and function.

The first aim of this book is to introduce the field to senior undergraduate
and postgraduate biologists. I have tried to show the ways in which our new
understanding of nucleocytoplasmic transport has come into being and in
what ways it is relevant to cell biology. I have assumed a basic general
knowledge of molecular and cell biology and of biochemistry, but no detailed
knowledge of any area. The first chapter places the field in its general context
and the second and third chapters provide the necessary background for

detailed discussion of important recent advances, which are the subject of Chapters 4 and 5. The sixth and final chapter surveys the actual and potential contribution of the field to cell biology. I have illustrated the early chapters fairly copiously but later ones more sparsely and Chapter 6 not at all; the intention is to facilitate basic understanding, but not to impose my own views on the reader where questions remain open.

The book's second aim is to use recent advances in the field to illustrate the process by which modern scientific progress occurs: the importance of developing techniques for formulating problems as well as providing answers to them, and the integration of different approaches leading to the evolution of novel concepts. This is why the longest chapter (Chapter 2), while ostensibly about methodology, is as much concerned with concepts and models; and Chapter 3, which is ostensibly about concepts, opens with a comment on techniques. In places, e.g. Sections 5.1 and 6.1, I have addressed relevant philosophy-of-science issues directly.

It is important to emphasize that this is a rapidly-developing research field. What is written today is likely to be superseded tomorrow. However, the general principles that I discuss in this book are unlikely to be rapidly outdated; only some particular details are likely to be changed (and new ones added) by the time the book appears in print.

No author can write on a subject of this kind without having drawn heavily on the ideas and findings of his colleagues throughout the world. This book took shape in the course of the numerous conversations that I have enjoyed with other workers in the nucleocytoplasmic transport field. I might single out the names of Carl Feldherr, Hugo Fasold, Bob Lanford, Gerd Maul, Werner Müller, Phil Paine, Oscar Pogo, Norbert Riedel, Joel Shaper and Walther van Venrooij because the total number of hours of discussion I have had with them probably exceed those that I have had with others. Those concepts that I am credited with originating, such as the solid-state transport model, grew directly from these discussions. But of course, I accept full personal responsibility for any errors that might be found in this book.

Acknowledgements

My warm thanks go to Ron Berezney, Gunter Blobel, Klaus Brash, Maria Carmo-Fonesca, Shona Comerford, Larry Gerace, William LeStourgeon, Gerd Maul, Werner Müller, Don Newmeyer, Phil Paine, Sheldon Penman, Alan Smith and Nigel Unwin for providing the photographs that are reprinted in these pages; and a particular word of thanks to Carl Feldherr for the electron micrograph of colloidal gold particle migration through the nuclear envelope that forms the cover of this book.

Thanks also to Jen Harvey for preparing the diagrams.

1 Introduction

The fundamental distinguishing feature of eukaryotic cells is the presence of a structurally and functionally separate nucleus. An obvious prerequisite for the cell's viability is the capacity to transport material in a selective and controlled manner between nucleus and cytoplasm. Nuclear proteins, such as histones and the nucleic acid polymerases, are synthesized in the cytoplasm and have to be conveyed to the nucleus, while specifically cytoplasmic proteins may have to be kept out. Mature messenger, transfer and ribosomal RNA molecules, together with their associated proteins, have to be exported from the nucleus to the cytoplasm, while their immature precursors have to be kept in. The mechanisms responsible for these highly specific and selective transport processes form the subject of this book.

1.1 BIOLOGICAL SIGNIFICANCE OF NUCLEOCYTOPLASMIC TRANSPORT

1.1.1 The control of nuclear functions

Over the past 20–30 years, biochemists and cell biologists have become increasingly interested in the ways in which key cellular activities such as DNA synthesis and protein biosynthesis are controlled. In many cases, the control of protein biosynthesis involves regulation of intranuclear functions – transcription and HnRNA processing – though cytoplasmic functions such as the control of mRNA stability are also important. The control of DNA and RNA metabolism within the nucleus involves specific proteins, many of which seem to enter from the cytoplasm at specific times. Understanding the mechanisms by which such proteins reach their sites of action in the nucleus when appropriate is therefore clearly important for understanding cellular control mechanisms. Also, many hormones exert their effects on target cells by modifying nuclear functions. Such hormone actions again raise questions about the transfer of material, or at least of signals, from the cell surface or the cytoplasm to the nucleus (Figure 1.1).

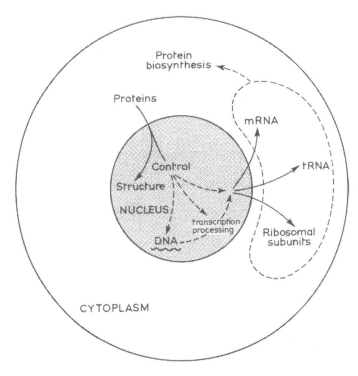

Figure 1.1 Schematic summary of the major aspects of macromolecule transport between nucleus and cytoplasm in relation to cell function.

1.1.2 Cell differentiation

Virtually all the cells in an adult multicellular eukaryote contain the same DNA, i.e. the same complement of genes. (There are exceptions: for instance, mature mammalian erythrocytes contain no nuclear material at all, and the antibody-secreting plasma cells derived from B lymphocytes contain some modified DNA. However, such exceptions are rare.) By and large, terminally differentiated cells vary as markedly as they do in appearance and function not because they *contain* different genes but because they *express* different genes. The globin genes are present in all nucleated cells in a mammal, but they are transcribed to a noticeable extent only in erythroblasts. Crystallin genes are likewise present in all nucleated cells, but they are transcribed to a significant extent only in the epithelial cells of the eye lens. Specific cytoplasmic proteins are once again important in determining which genes are to be transcribed and which are to remain silent, and therefore in determining the composition, structure and function of the cell. These proteins not only have to enter the nucleus in order to act, they also have to enter at appropriate times in the cell's and the organism's life.

A classic experiment (Gurdon, 1970), simple to describe though not simple to perform, illustrates both the constancy of DNA in all cell types and the role of the cytoplasm in selecting genes for transcription (Figure 1.2). Amphibian erythrocyte nuclei are very small, have highly condensed chromatin, and are devoid of activities such as DNA replication and transcription. Amphibian oocytes are very large cells, suitable for experiments

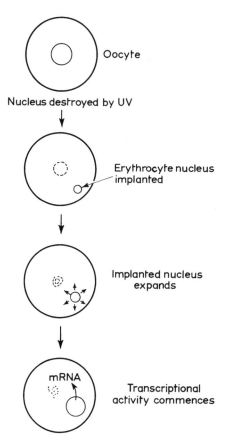

Figure 1.2 The nuclear transplantation experiment.

involving microinjections and other manipulations. If the nucleus of an amphibian oocyte is removed or destroyed by ultraviolet irradiation, and the nucleus of an erythrocyte from the same species is implanted in its place, a remarkable thing happens (when the experiment works, that is: the 1–2% of these experiments that succeed are important; the 98–99% that fail are scientifically irrelevant). The implanted nucleus swells up to perhaps 200 times its initial volume. Its chromatin decondenses, and transcription begins. Most importantly, the genes that are transcribed do not include globin genes, as we might expect of a nucleus from an erythrocyte; rather, they are genes coding for proteins characteristic of oocytes. It is the cytoplasm, not the nucleus, that determines which genes are expressed. An understanding of how the proteins involved in this effect enter the nucleus and act (Dreyer *et al.*, 1983) is basic to our understanding of the mechanisms of cellular differentiation.

1.1.3 RNA transport and the control of protein biosynthesis

So far, we have focused on the biological significance of protein import to the nucleus. What about the significance of nucleic acid export?

Protein biosynthesis can be controlled at many levels. Transcription in the nucleus, and the control of mRNA half-lives in the cytoplasm, seem to be the two most generally important levels in most eukaryotic cells. However, intermediate stages in mRNA metabolism – RNA processing, nucleo-cytoplasmic transport and translation – are not irrelevant. For instance, there is reasonable evidence that an early effect of many carcinogens is a disturbance of normal mRNA transport from nucleus to cytoplasm, resulting in an altered cytoplasmic messenger population, sometimes with no concomitant change in transcription (Shearer, 1974). In some cell types, mRNA transport, as opposed to transcription, processing, translation and the regulation of half-life, seems to be a target for hormonal regulation (Goldfine *et al.*, 1982).

According to our present understanding (Chapter 5), mRNA transport is a biochemically complicated process, involving the kind of elaborate mechanism that we often find in controllable processes in biology. The mechanisms of tRNA and snRNA transport are also biochemically interesting, although it is not yet clear whether they have much significance in the control of cellular activities. The mechanism of ribosome transport remains obscure; it is the focus of research at present.

1.2 THE STUDY OF NUCLEOCYTOPLASMIC TRANSPORT

This brief overview shows the significance of an understanding of nucleo-cytoplasmic transport in cell biology as a whole. The general issue of biological significance will be discussed in Chapter 6. However, the study of this field has another advantage from the student's point of view. Much of our present knowledge in the field has come from research carried out over the past 5–10 years, deploying a wide range of modern techniques. Therefore, a survey of the subject as it stands today illustrates clearly the ways in which (for instance) recombinant DNA technology, monoclonal antibodies, and new techniques in microscopy, cell manipulation, protein chemistry and nucleic acid chemistry have been brought to bear synergistically on both the formulation and the resolution of a particular set of research problems. It is a topical example of the way in which different disciplines can contribute harmoniously to progress in science. More generally, this field provides a good illustration of the ways in which conceptual and methodological advances are integrated in an area of scientific research.

1.3 SOME GENERAL CONSIDERATIONS

1.3.1 What is 'transport'?

In cell biology, we usually think of 'transport' as the movement of a solute (e.g. a small metabolite such as glucose, or an ion such as Na^+) from one aqueous compartment to another across a non-aqueous barrier such as a membrane. But the barrier separating the nucleus from the cytoplasm, the **nuclear envelope**, is not a simple membrane. It contains non-membranous structures, the **pore complexes**, which permit rapid exchange of virtually all except high molecular weight solutes. Small solute molecules, including globular proteins with molecular weights up to about 15 000, diffuse through the nuclear envelope as quickly as they do in the cytoplasm (Fenichel and Horowitz, 1969; Paine and Horowitz, 1980). In other words, the nuclear envelope has the properties of a molecular sieve. This is why only the transport of macromolecules – proteins and nucleic acids – has been mentioned in the previous sections.

So far as we know, all solutes pass from nucleus to cytoplasm or from cytoplasm to nucleus via the pore complexes; no alternative route has been established. In principle, all three classic membrane transport mechanisms – passive diffusion, facilitated diffusion and active transport – might apply to macromolecule movement through these structures, but the exchange of small molecules and ions is almost certainly a matter of passive diffusion only. However, proteins and nucleic acids are not always freely mobile solutes *in vivo*. In fact, as time goes on, there is increasing evidence that more and more of these molecules are effectively immobolized *in situ* by adsorption or specific binding to insoluble structures within the cell – membranes or cytoskeletal elements, for example. This means that the approach to

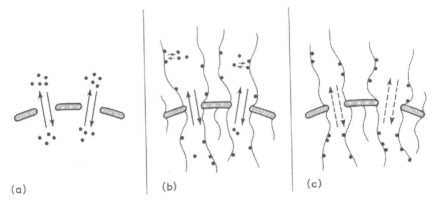

Figure 1.3   Three general schemes representing possible nucleocytoplasmic transport mechanisms. (a) The 'solution-diffusion model', assumed for conventional membrane transport processes and presumably relevant to nucleocytoplasmic exchanges of, for example inert microinjected materials such as colloidal gold. (b) A modification of this scheme to take account of cytoplasmic and intranuclear binding (partial immobilization) of the transportable macromolecule. (c) The solid-state model: migration of the macromolecule between the compartments depends on either movement along fibrils, or movement of the fibrils themselves.

nucleocytoplasmic transport cannot necessarily be confined to a study of the activities of the nuclear envelope, important as these are.

Figure 1.3 illustrates the point schematically. Figure 1.3(a) shows the movement of a macromolecule that is freely soluble in both nucleoplasm and cytoplasm through the pore complexes. This model is analogous to 'classic' membrane transport processes, and could involve carriers or energy transduction or passive permeation. Figure 1.3(b) illustrates the movement of a macromolecule that is in equilibrium between bound and freely-soluble states in both nucleoplasm and cytoplasm; the soluble form can pass through the pore complexes. According to this model, transport from cytoplasm to nucleus involves: (i) desorption from the cytoplasmic binding sites; (ii) movement through the pore complexes, which may once again be carrier-mediated or energy-requiring or merely passive; and (iii) attachment to the intranuclear binding sites. It goes somewhat against intuition to broaden our understanding of 'transport' to include events that are spatially and temporally remote from the obvious 'diffusion barrier', but the point is not merely academic; it is now fairly clear that protein transport from nucleus to cytoplasm often involves just such 'remote' events (Chapters 2 and 4).

Figure 1.3(c) is a more radical extension of this model. Here, the macromolecule is never freely soluble or diffusible, but exists in all compartments – nucleoplasm, envelope and cytoplasm – only in a tightly-bound form. Once again, transport involves minimally three stages: **release** from the intranuclear binding sites, **translocation** through the pore complexes, and **cytoplasmic** (or **cytoskeletal**) binding. Once again, the second of these stages – translocation – can involve facilitated diffusion or active transport (though passive diffusion is obviously not a possibility in this case). And once again, this **solid-state transport** model is no mere abstraction. All currently available evidence leads to the conclusion that nucleocytoplasmic transport of mRNA occurs according to just such a scheme (Agutter, 1985a and Chapter 5). If it has achieved nothing else, the study of nucleocytoplasmic mRNA transport has added the concept of solid-state transport to the conceptual repertoire of cell biology.

1.3.2 The nuclear envelope

Whichever of these general models of transport we adopt in any particular case, the structure and function of the nuclear envelope are clearly relevant issues. The nuclear envelope has five ultrastructurally distinct components:

1. the **outer nuclear membrane** (ONM), which is similar in composition and function to the rough endoplasmic reticulum (ER) and in many cells may be continuous with this membrane system (Richardson and Agutter, 1980);
2. the **inner nuclear membrane** (INM), which is different in composition and physical properties, and probably different in function, from the ONM;
3. the **perinuclear cisterna**, an enclosed space between the ONM and INM, separated from both nucleoplasm and cytoplasm by these membranes but perhaps continuous with the cisterna of the ER in cells where the ONM and ER are continuous;
4. the **pore complexes** (PCs);
5. the **lamina**, a fibrillar protein network tightly linked to the PCs and to the nucleoplasmic face of the INM and bearing much the same relationship to the INM as the spectrin–actin framework does to the erythrocyte membrane.

Biochemical studies have given us a fairly detailed knowledge of the structure and function of the ONM and the lamina, and some insights into the nature of the INM; but they have only recently begun to give us an understanding of the composition and organization of the all-important PCs. Because of this and of the numerous electron-microscopic studies of the PC, at least 27 different models of PC organization, all based exclusively on ultrastructural and biophysical evidence, have been discussed in the literature (Archega and Bahr, 1985). Figure 1.4 shows some of these models and in each case illustrates the relationship of the PC to ONM, INM, cisterna and lamina.

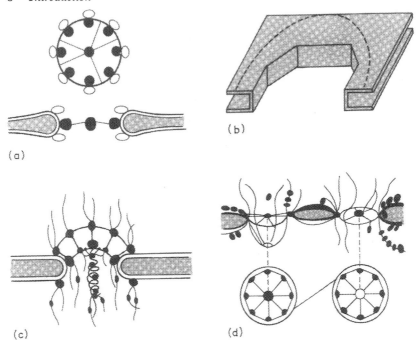

(a)

(b)

(c)

(d)

Figure 1.4   Four of the many attempts to interpret PC organization on the basis of ultrastructural evidence. For further discussion see Franke, 1974; Unwin and Milligan, 1982; Archega and Bahr, 1985; and Agutter, 1988.

Those aspects of nuclear envelope biochemistry that are relevant to nucleocytoplasmic transport, including the recent analyses of the PC, will be discussed in Chapter 3.

1.3.3 Other structures

In so far as nucleoplasmic and cytoplasmic binding and release events are relevant to macromolecule transport between the compartments (cf. Figure 1.3), the structures responsible for binding and the mechanisms of binding and release are relevant. What are these structures?

Within the cytoplasm, the cytoplasmic membranes and the various elements of the cytoskeleton are obvious candidates. Membrane systems such as the ER can present very large surface areas in some cell types and even if the affinity of a protein for such a membrane is quite low, the majority of the protein can therefore be adsorbed to the membrane *in situ*. The cytoskeleton – microtubules, microfilaments and intermediate filaments — can also present a large total intracellular surface area and might contain specific high-affinity binding sites for particular proteins, making such proteins effectively non-diffusible. There is evidence, too, that all the translationally active (polysomal) mRNA is bound to the cytoskeleton (van Venrooij *et al.*, 1981); the term 'cytoskeleton' here includes the immobile protein network associated with the phospholipid bilayer of the rough ER (Capco and Penman, 1983). Evidence relating to the importance of these sites of macromolecule binding will be discussed in Chapters 2 and 5.

Within the nucleus, the chromatin in general, and DNA in particular, are the most obvious candidates. Proteins that accumulate in the nucleus (**karyophilic** proteins) often do so because they contain 'signals' that permit translocation through the PCs and also domains for binding to intranuclear components, such as particular DNA sequences. The growth of interest in specific DNA binding proteins over the last few years has promoted the growth of understanding of this issue; details will be discussed in Chapter 4.

It is generally (though not universally) believed that nuclei contain a structure commonly known as the **nuclear matrix** or **nucleoskeleton**, which bears roughly the same relationship to the nucleoplasm as the cytoskeleton does to the cytoplasm. The actual nature of the nucleoskeleton is controversial at present; some aspects of the controversy that are relevant to nucleocytoplasmic transport will be discussed in Chapters 2 and 3. However, the main point is this: the existence of some kind of nucleoskeleton implies that there is another candidate for the provision of intranuclear macro-molecule binding sites. It seems that mRNA and its precursors within the nucleus are attached to just such sites. In some systems at least, the same may well be true of ribosomes and their precursors (Chapter 3).

1.4 OVERVIEW

The points made in this chapter suggest that three kinds of experimental study can throw light on nucleocytoplasmic transport mechanisms: (a) detailed chemical and structural studies on the transportable macromolecules them-selves, which reveal internal characteristics that make them transportable; (b) biochemical and ultrastructural analyses of not only the nuclear envelope, but also cytoplasmic and nucleoplasmic structures that might be involved in macromolecule binding; and (c) actual measurements of the rate of transport and extent of accumulation of macromolecules between and within compart-ments, using *in vivo* and *in vitro* methods. The second and third of these experimental approaches form the topics of the next chapter. The first group involves methods that are common to most areas of modern biochemical and molecular biological research; they have been described in several widely-available textbooks, so an account here would be redundant.

The elucidation of nucleocytoplasmic transport mechanisms by these three methodological approaches is relevant to our understanding of several major topics in cell biology: the maintenance of composition, structure and function of the nucleus and cytoplasm; the regulation of protein biosynthesis; and the control of cell differentiation.

2   Methods of study

When the first electron-microscopic studies of the nuclear envelope (NE) were carried out in 1949–50 (Callan *et al.*, 1949), they raised an obvious question: how widespread in nature were the most striking features of the envelope, the PCs? Further electron-microscopic studies carried out during the 1950s led to the following general answers.

1. All NEs have PCs.
2. All PCs, irrespective of species and cell-type of origin, have identical (or very similar) shapes and sizes.
3. The numbers of PCs per nucleus vary enormously with the type and metabolic state of the cell.
4. Apart from the NE, only one type of membrane system, known as annulate lamellae, has PCs. Annulate lamellae are found most commonly in rapidly-dividing cells such as those of early embryos, and are generally believed to be pools of NE ready for incorporation after some future mitosis.

Apart from the inevitable debates about choice of fixing and staining procedures and the danger of artefacts in electron microscopy, these findings led to speculations about the biological functions of PCs. (For an excellent early review, see Gall, 1964.) The most popular hypothesis, that they acted as channels for nucleocytoplasmic exchange, was first tested empirically by Feldherr (1962). He injected colloidal gold coated with polyvinylpyrrolidone into the cytoplasm of *Amoeba proteus*, and examined the cells by electron microscopy at various times after injection. The results were unequivocal: equilibration of the colloidal particles in the cytoplasmic compartment was rapid; after a delay, significant accumulations appeared in and around the

NE. Subsequently, over a time-course of several hours, the particles penetrated through the PCs and became visible in the nucleoplasm.

The hypothesis that the PCs are transport channels gained further support from a study by Stevens and Swift (1966) on the transcripts of the lampbrush chromosomes of *Chironomus* salivary glands. These giant polycistronic messengers, which code for the insect's salivary glycoproteins, can readily be seen in the electron microscope, and Stevens and Swift obtained micrographs showing the messengers apparently in transit through the PCs.

These publications by Feldherr and by Stevens and Swift were landmarks in the field. The techniques described in them — microinjection of cells with identifiable permeants, and analysis of transport events by microscopy (especially electron microscopy) — have been fundamental to the growth of our knowledge of nucleocytoplasmic transport.

2.1 *IN SITU* PERMEABILITY STUDIES

2.1.1 Permeability properties of the envelope

Subsequent studies by Feldherr and his colleagues showed that the rate of entry into the nucleus of both colloidal gold particles (Feldherr, 1972) and fluorescently-labelled proteins (Paine and Feldherr, 1972) after injection into the cytoplasms of large cells was inversely related to particle size (or molecular weight) (Figure 2.1). The colloidal gold used in such experiments was prepared by mixing ether-phosphorus with an aqueous solution of auric chloride, the range of particle sizes being a function of the relative volumes of the two solutions. Coating with polyvinylpyrrolidine confers an inert hydrophilic surface on the particles, making them suitable for microinjection studies; moreover, the particles are electron-dense and therefore well suited for electron-microscopic observation. Movements of the fluorescently-labelled proteins were followed by fluorescence microscopy. (In many subsequent protein transport experiments, tritium-radiolabelled rather than fluorescently-labelled proteins have been used, the distribution of the label after various times usually being determined by electron-microscope autoradiography. Tritium is more suitable for autoradiographic work than any other radioisotope because it gives better spatial resolution: the mean pathlength of the β-particles is shorter even than that of ^{14}C.) In all cases, the findings were consistent with earlier results: the NE corresponded structurally to the permeability barrier between nucleus and cytoplasm, and the channels through which permeation occurred were the PCs (Figure 2.2).

In the light of these qualitative studies, the obvious problem in the early/mid 1970s was to obtain a quantitative estimate of the functional diameter of the permeation channels in the PCs. This problem was solved by Paine *et al.*

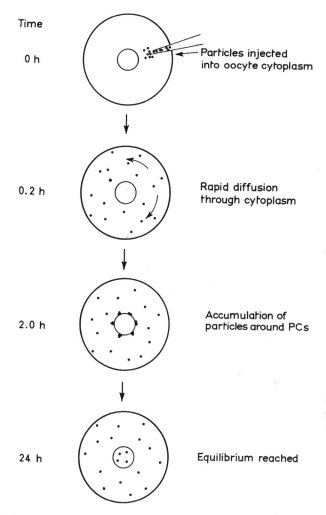

Figure 2.1 Schematic representation of the course of a microinjection experiment. Visualizable particles are injected into the cytoplasm at time Oh; the processes of accumulation around the nuclear envelope and entry into the nucleus are more or less retarded depending on the median particle size.

(1975). They used tritium-labelled dextrans, which have the advantages of being more or less spherical particles, biologically inert, stable within the cell, and water-soluble; and they obtained narrow and well-quantified ranges of dextran size by gel filtration. Each dextran preparation was injected into the cytoplasm of an amphibian oocyte, and the time-course of entry into the nucleus was determined by autoradiography. The calculation of the patent

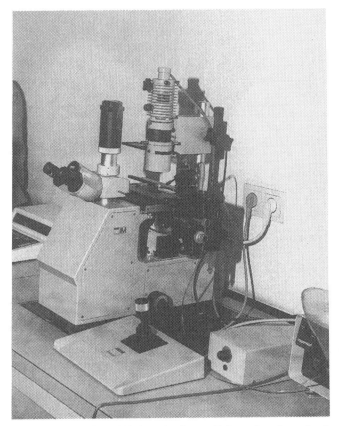

Figure 2.2 Equipment for studying the intracellular migration of microinjected fluorophores: a Zeiss immunofluorescence microscope fitted with an Eppendorf Microinjector 5242. Photograph kindly taken, at the author's request, by Prof. Dr W. E. G. Müller.

pore radius from the results was mathematically rather complicated and involved the introduction of certain assumptions, such as a value for the length of the PC channel which was based on electron-microscopic studies of thin sections of nuclei. Nevertheless, mutually consistent results were obtained with dextrans of markedly different sizes: the mean value for the patent (functional) radius of the PC was about 4.5 nm.

This value was immediately credible for two reasons. First, it was consistent with the results of previous permeability studies (see above) which suggested that the NE was freely permeable to inorganic ions and other small solutes and effectively impermeable to proteins and other particles larger than bovine serum albumin, around 67 kDa. ('Effectively impermeable' means that diffusion at or above this size range is extremely slow.) Second, it corresponded to the size of the 'central granule' or 'central spot' observed in many electron micrographs of PCs, and thus lent support to the view that this 'central granule' was in fact the functional communication channel through the NE (Abelson and Smith, 1970). Nevertheless, perhaps because of the complexity of the calculations and the number of assumptions required, there was no comparable investigation for almost a decade; and the fact that amphibian oocytes are in many respects atypical cells – not only because of their enormous size – left room for doubt that the 4.5 nm measurement was valid for all cell types.

The technique of laser photolysis (photobleaching), introduced to the field by Peters and his colleagues in the 1980s, has allowed comparable measurements to be made on much smaller cells (Lang and Peters, 1984). In essence, the technique is as follows. A fluorescently-labelled particle (e.g. a dextran or a protein) is injected into the cytoplasm of the cell under study and allowed to diffuse to equilibrium. It is important that the fluorophore can be irreversibly bleached by a suitable laser, and that it is irreversibly bound to the particle. Because most cells are thickest in the region of their nuclei, the fluorescence profile at equilibrium is typically that shown in Figure 2.3(a). The region containing the nucleus is then irradiated by the laser for about 1 ms; this destroys the fluorescence in the nucleus, but the fluorescence in most of the cytoplasm survives (Figure 2.3(b)). The time-course of recovery of fluorescence in the region of the nucleus is then measured; since the bleaching is irreversible, this recovery must represent the diffusion of fluorescently-labelled particles from the cytoplasm to the nucleus (Figure 2.3(c)). The applicability of this procedure to small cells stems from the high spatial resolution (around 1 μm) attainable by fluorescence microscopy, and the high time resolution (around 1 ms) consequent on the rapidity and completeness of bleaching by the laser. These resolutions are far better even than those attainable by tritium autoradiography, where the number and size of grains on the exposed X-ray plate is an important factor. Peters and his colleagues obtained measurements of patent pore radii in cells such as hepatocytes and HeLa cells, which are commonly used in cytological and biochemical research, and these measurements are reassuringly compatible with those obtained by Paine *et al.* (1975) for the amphibian oocyte: around 4.5–6.0 nm. Despite the agreement, there is a lingering doubt about studies of this kind on small cells. Injection of (say) 0.1 picolitres of material into an amphibian oocyte makes no significant difference to the cell volume or presumably to the intracellular organization; this is not true of a hepatocyte. Even the relative diameter of the microinjection needle compared to the cell's dimensions might constitute an important methodological difference.

However, this 4.5–6.0 nm value is generally accepted and quantifies the most important general permeability property of the nuclear envelope: its

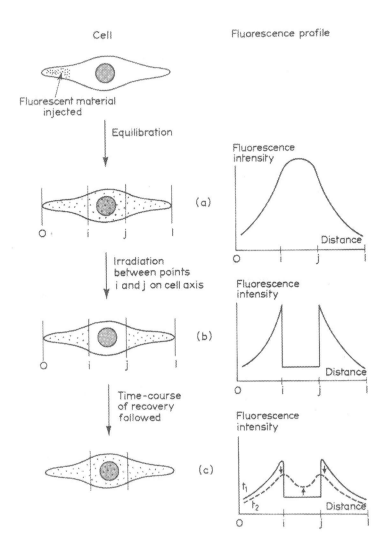

Figure 2.3 Schematic representation of the course of a laser photolysis experiment. A section through the cell under study (left) is shown alongside the corresponding fluorescence profile (right). Some time after microinjection, the fluorophore is distributed uniformly along the cell axis (O–l). Because the cell is usually thickest in the region of the nucleus (i–j), the fluorescence intensity is greatest over this part of the axis (Figure 2.3a). The portion i–j is then irradiated for about 1 ms with a suitable laser and the fluorescence in this region is bleached (Figure 2.3b). The time-course of recovery of the fluorescence in this region is then determined (Figure 2.3c); this is largely determined by the rate of diffusion of the surviving fluorophore into the nucleus.

action as a molecular sieve. It provides a basis for predicting the passive diffusion rate across the nucleo-cytoplasmic boundary of any permeant of which the *in situ* diameter can be reliably estimated.

2.1.2 Experiments on specific transport processes

Studies aimed at measuring the patent pore size necessarily make use of permeants for which the cell has no specific transport machinery. Obviously, any such machinery invalidates the assumptions pertinent to passive diffusion processes. Non-biological particles of macromolecular size (dextrans and colloidal gold) or proteins foreign to the cell are therefore appropriate materials for such work.

Conversely, measurements of the patent pore radius tell us little or nothing useful about the nucleocytoplasmic movement of any macromolecule for which there *is* specific transport machinery. Even in the early 1970s there were numerous lines of evidence to indicate that specific transport mechanisms existed. For instance, the mere fact that some nuclear (and obviously highly karyophilic) protein subunits, such as the RNA polymerase core polypeptides, had molecular weights greatly in excess of the passive diffusion limit of the PC meant either that there were specific transport mechanisms or that such polypeptides entered the nuclear compartment only when the NE broke down at mitosis; and the latter possibility was excluded by cell-cycle studies, and was in any case inconsistent with the maintenance of an intact NE through mitosis in some cell types such as unicellular fungi. In addition, several studies (e.g. Jelinek and Goldstein, 1973) showed an extremely rapid shuttling of certain proteins – and small RNA species – between the nuclei and cytoplasm; in heterokaryons, some molecules seemed to spend only very brief times in the intervening cytoplasm (Fig. 2.4). In experimental studies of specific transport mechanisms nowadays, it is normally necessary to microinject the purified substance or to permeabilize the cell for its entry, and then to use a detection method that is reliably specific. It is also important to ensure that the permeant is not degraded (or otherwise seriously modified) during the course of the experiment; unfortunately, this criterion is not always satisfied.

Many experiments of this type have been performed with techniques very similar to those described in the previous subsection, e.g. using permeants labelled with tritium or a fluorophore. The study by Yamaizumi *et al.* (1978) showing rapid uptake of cytoplasmically microinjected non-histone proteins into nuclei was of this kind. So were the more recent studies by Zasloff (1983) on tRNA export from nuclei; these latter experiments involved microinjection into the nuclei rather than the cytoplasm of amphibian oocytes. Other experiments have involved colloidal gold particles coated with the permeant under investigation instead of polyvinylpyrrolidine; a good example is the

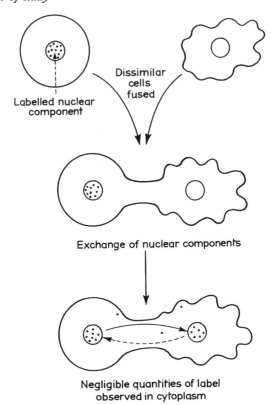

Labelled nuclear
component

Dissimilar
cells
fused

Exchange of nuclear components

Negligible quantities of label
observed in cytoplasm

Figure 2.4 Exchange of material between the nuclei in a heterokaryon. Such demonstrations that karyophilic materials could in fact shuttle between nucleus and cytoplasm originated with the studies by Goldstein and his colleagues in the early 1970s.

Figure 2.5 Summary of the kinds of experiments involved in exploring the signal sequences characteristic of karyophilic proteins. Typically, a monoclonal antibody against the protein is required and the gene must be isolated and cloned, both in wild-type and variant forms (especially variants with key portions of the sequence removed or replaced). From the cloned genes, a family of wild-type and mutant proteins is generated. Each protein is used in a microinjection experiment and its nucleo-cytoplasmic distribution is determined by immunofluorescence microscopy, using the monoclonal antibody). It is important in such experiments to demonstrate by Western blotting that neither the wild-type nor any of the variant proteins are degraded within the cell. The evidence from such experiments is necessary but not quite sufficient for unequivocal identification of a signal sequence (see Chapter 4 for further discussion).

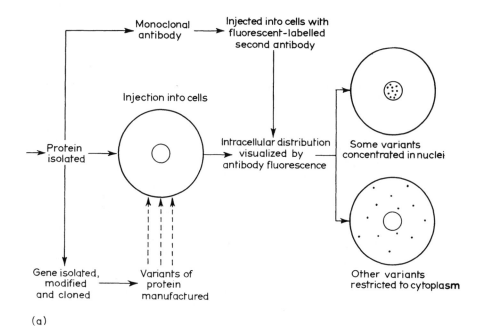

(a)

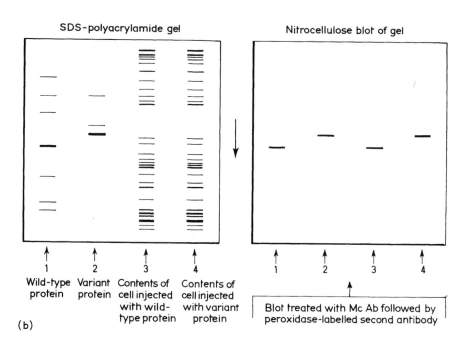

(b)

study by Feldherr *et al.* (1984) using nucleoplasmin-coated gold. However, an increasing majority of studies on nucleocytoplasmic protein transport exploit the rapid growth in availability of monoclonal antibodies. Typically, the unlabelled protein is injected into the cytoplasm (more rarely into the nucleus) of the cell under study, and after an appropriate time, which may vary from 1 min to 24 h, the cell is permeabilized to allow entry of a monoclonal antibody against the protein and of a fluorescently-labelled second antibody (Figure 2.5(a)). The nucleocytoplasmic distribution is then demonstrated by fluorescence micrographs. (The comments made earlier about the sizes of microinjection needles and the relationship between the volumes of the cell contents and the injected material apply to many of these experiments. Also, the danger of redistribution of contents after permeabilization has to be considered.) In principle, though rarely in practice, the monoclonal antibody can also be used on Western blots (Figure 2.5(b)) to show: (i) that the protein remains undegraded during the time-course of the experiment, and (ii) that the antibody is really specific for the protein. Without this Western blotting evidence, fluorescence micrographs cannot always be interpreted unequivocally. The most widespread use of such experiments at present is on proteins that have been modified by genetic manipulation, the usual objective being to identify regions in the wild-type protein that serve as signals in nucleocytoplasmic transport. Several examples will be discussed in Chapter 4.

Such techniques have been responsible for most of our recently-acquired knowledge about nucleocytoplasmic transport mechanisms, especially those concerning proteins.

2.1.3 Which macromolecules are diffusible *in situ*?

In the years around 1970, advances in electron microscopy and the development of techniques such as immunofluorescence revealed new information which transformed the traditional biochemist's image of the cell. Intracellular processes could no longer be envisaged, even approximately, as taking place in an unstructured aqueous solution. During the 1970s and early 1980s the alternative image of the cell generated by these advances became steadily more entrenched: a highly structured environment in which most of the water and most of the 'solutes' are more or less bound. The initially controversial account of a cytoplasmic 'microtrabecular lattice' by Wolosewick and Porter (1976) and a more recent and much-cited essay by Fulton (1983) were symptomatic of this change in perspective.

Two series of experiments in the early 1980s were crucial in adapting our understanding of nucleocytoplasmic transport processes to the altered perspective. Feldherr and his colleagues considered two alternative classes of explanations for the high nucleus : cytoplasm concentration ratios *in situ* of karyophilic proteins and of nucleus-restricted RNAs: (i) there were specific (active) transport mechanisms in the NE capable of concentrating materials in the nucleus; (ii) there were intranuclear binding sites for the materials in question. (These two classes of explanation are not mutually exclusive, of course. See, for example, Feldherr and Pomerantz, 1978.) Paine and his colleagues were concerned by the generally poor correlation between protein molecular weight and nucleus:cytoplasm concentration ratio. (Obviously if passive diffusion is the principal mechanism for cytoplasm-to-nucleus protein movement, a good correlation would be expected. See, for example, Austerberry and Paine, 1982.) They addressed the question: how many of these discrepancies can be explained in terms of the immobilization of proteins in the cytoplasmic compartment (or, conversely, do the solid-phase parts of the crowded cytoplasm affect nucleus:cytoplasm concentration ratios by excluding solutes just as gel-filtration chromatography materials do?) These studies in Feldherr's and Paine's laboratories revealed that binding of macromolecules is extensive within both the cytoplasm and the nucleus. In both cases, the cell type chosen for study was again the amphibian oocyte; the size and ready availability of these cells, and the substantial history of experiments on nucleocytoplasmic transport that have involved them, easily account for the choice.

Feldherr's approach was simple (Figure 2.6). Inserting a microdissection needle into the cell (Figure 2.6), he tore the NE *in situ*. After this insult, the plasma membrane resealed but the NE did not; therefore there was no longer a permeability barrier between nucleus and cytoplasm, so if a high nucleus:cytoplasm concentration ratio of any material survived the injury it

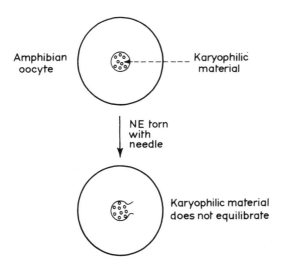

Figure 2.6   Feldherr's nuclear envelope puncturing experiment.

had to imply that that material was bound within the nucleus. The cells were left for various times, then they were microdissected and the compositions of cytoplasm and nucleus were examined. Although the nucleocytoplasmic distributions of soluble, diffusible proteins were rapidly altered after the NE was mutilated (equilibration between the compartments was observed, as expected), neither the known karyophilic proteins nor the rRNAs had their distributions changed significantly. Feldherr and his colleagues concluded that high intranuclear concentrations of macromolecules could be accounted for by intranuclear binding; they were unable to comment on the significance of NE transport processes, but their work left the implication that such processes were probably not important in most cases. As for the nature of the intranuclear binding sites, the controversial 'nuclear matrix' was presented as a possibility.

More recently, Dreyer *et al.* (1986) have performed even more drastic experiments, stripping the entire NE from the oocyte nucleus *in situ*. They found, in contrast to Feldherr's group, that the nucleus-bound components equilibrated with the cytoplasmic compartment over intervals of 15–20 min, and concluded: (i) that intranuclear binding was not important in determining nucleocytoplasmic distributions, as Feldherr had claimed, (ii) that there was no evidence for an *in situ* 'nuclear matrix'. These inferences may be overstated. For instance, the integrity of 'matrix' or other intranuclear binding sites might be dependent on anchoring to the NE; cytoplasmic proteinases might have entered the flayed nuclei and destroyed the binding sites; and the equilibration time might, if subjected to detailed biophysical analysis, reveal that fairly high-affinity binding sites survived the injury after all. However, they suggest that some difficulties of interpretation surround Feldherr's observations. Neither Dreyer nor anyone else would doubt, for example, that the approximately 500:1 nucleus:cytoplasm concentration ratio of histones is largely a consequence of the well-characterized binding of these proteins to DNA; but perhaps some other high intranuclear concentrations have less to do with binding than Feldherr inferred from his results.

The studies by Paine and his colleagues, involving the 'reference phase' technique (Figure 2.7), were initiated around 1981 and exploited some then-recent advances in polyacrylamide gel electrophoresis of proteins: two-dimensional separations, which resolve on the basis of isoelectric point as well as molecular weight; and the highly sensitive silver staining method. Briefly, an aqueous gelatin solution (the volume was roughly equal to the volume of the nucleus) was injected into the oocyte cytoplasm (Figure 2.7). Any diffusible protein in the cytoplasm would migrate freely into the gelatin sphere or 'internal reference phase'; non-diffusible proteins could not migrate, of course. After time for equilibration, the cells were cryodissected, i.e. they were flash-frozen in liquid nitrogen and the nucleus, reference phase and bulk cytoplasm were separated from each other by microdissection. Each of these three compartments was then subjected to two-dimensional electrophoresis and the proteins in them were revealed by silver staining (Figure 2.8(a), (b)). The results justified a number of striking and largely unexpected conclusions. Despite the obvious criticisms of these experiments: (i) the injection of a relatively large volume of gelatin might perturb the organization of the cytoplasm significantly, and (ii) because the reference phase gel excludes very large molecules it might distort the profile of the diffusible proteins — the results remain valid. If the first criticism is sound, the effect will be to diminish rather than to increase the number of

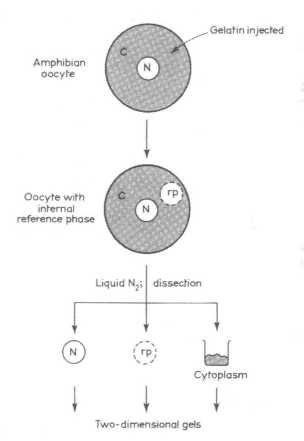

Figure 2.7 The internal reference phase/cryodissection procedure used by Paine and his colleagues. Diffusible proteins in the cytoplasm (C) can enter the reference phase (rp) which is provided by the microinjected gelatin. After time for equilibration, the cell is flash-frozen in liquid nitrogen and dissected while frozen into nucleus (N), cytoplasm and reference phase. Because of the sensitivity of the silver staining procedure and the large size of the amphibian oocyte, one cell can provide sufficient material for an adequately-resolved two-dimensional gel.

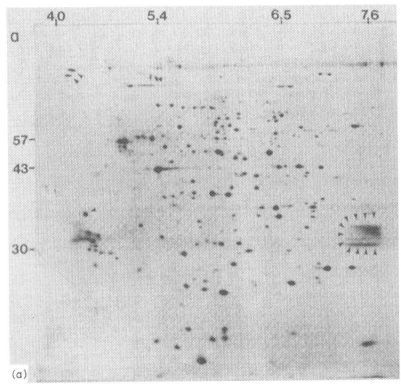

Figure 2.8 The two-dimension gels of the cryodissected (a) reference phase and

non-diffusible proteins. The second criticism is irrelevant to the movement of proteins into the nucleus by diffusion: proteins that are too big to enter the reference phase are too big to pass through the channels in the pores at significant rates by passive means.

The conclusions were as follows. First, it was apparent that about one-third of the cytoplasm's proteins were non-diffusible; the diffusibility of many of the others was restricted by, for example, gel exclusion effects. Therefore, protein binding within the cytoplasm was an important determinant of nucleocytoplasmic distribution for some proteins at least. Secondly, although there was some qualitative correspondence between the lower molecular weight proteins in the reference phase and those in the nucleus, indicating that the nucleocytoplasmic distributions of several such proteins might indeed result from passive diffusion, there were many exceptions. These made it possible to identify those proteins for which the oocyte had

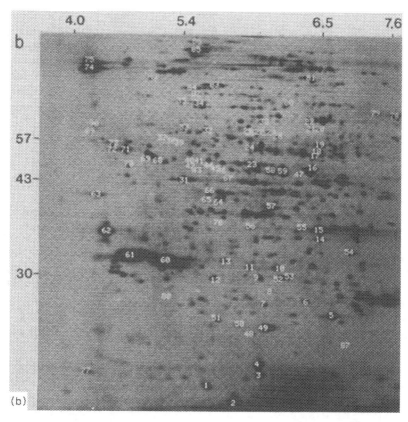

(b) nucleus from an oocyte, as described in Figure 2.7. The horizontal axes are pI values, the vertical axes molecular weights (kDa). Photographs kindly supplied by Dr P. L. Paine and reproduced by permission of the Editorial Board of *J. Cell Biol.*

specific mechanisms for maintaining high nuclear concentrations. Thirdly, and perhaps most surprisingly, the protein composition of the nuclei was markedly different from the composition of nuclei isolated by conventional cell-fractionation techniques, and even by conventional oocyte microdissection (as opposed to cryodissection).

A dramatic illustration of the difference was provided by nucleoplasmin, a large acidic pentameric protein which is believed to function in nucleosome assembly (it apparently 'presents' the newly-synthesized histones H2A and H2B to the DNA at the end of the cell's S-phase; the other core histones seem to be 'presented' by the nuclear proteins N1 and N2). In nuclei isolated by fractionation of aqueous oocyte cell homogenates, nucleoplasmin can scarcely be detected; but Paine and his colleagues (1975) showed that it represented almost 30% of the total nuclear protein *in situ*. The inference must be that nucleoplasmin and other important nuclear proteins are lost, in whole or in part, during the preparation of nuclei from homogenates. If this is true even for the huge nuclei of amphibian oocytes, the loss of such proteins must be far more rapid and complete for ordinary-sized nuclei such as those of hepatocytes.

This finding obviously has serious implications for experiments performed with isolated nuclei.

2.2 *IN VITRO* TRANSPORT STUDIES

2.2.1 Experiments with isolated nuclei

Protein depletion occurs even in oocyte nuclei isolated by gentle micro-dissection techniques. When nuclei are isolated in bulk, e.g. from liver homogenates, they are not only seriously protein-depleted but also mechanic-ally damaged: the envelopes of more or less all bulk-isolated nuclei are fairly badly torn. This is indicated by the rapid entry into them of large proteins and protein–dextran complexes (Burgoyne and Skinner, 1979) as well as by ultrastructural evidence (Figure 2.9). On the face of it, it is highly improb-able that an organelle that has suffered an unknown amount of functional impairment (what were the *in vivo* functions of all the lost proteins?) and has large breaks in its surface permeability barrier could exhibit any physiologic-ally relevant transport properties.

To make matters worse, cell or tissue homogenization liberates proteinases that attack the nuclear surface; protein degradation is a serious problem in the study of most subnuclear fractions and certainly affects isolated nuclei (Kaufmann *et al.*, 1981). Tissues such as liver, from which relatively good nuclear preparations can be obtained (yields are high and visible cytoplasmic contamination is low), are particularly rich in proteinases. Many of these are serine proteinases and can be inhibited with reagents such as phenylmethane-sulphonyl fluoride (PMSF), but 'many' is not 'all'. Artefactual protein–protein cross-links also form in nuclei during isolation, for example as a result of sulphydryl group oxidation (Kaufmann *et al.*, 1981). Finally, isolated nuclei, even 'highly purified' preparations, are invariably contaminated with at least some cytoplasmic material – collapse of the perinuclear intermediate filament bundles on to the nuclear surface during homogenization is inevitable.

Several of these difficulties can be avoided by isolating nuclei in non-aqueous media, but nuclei isolated in this way are of very limited value. Any functional component of a NE transport system is likely to be irreversibly inactivated. In any case, a realistic *in vitro* transport experiment must involve

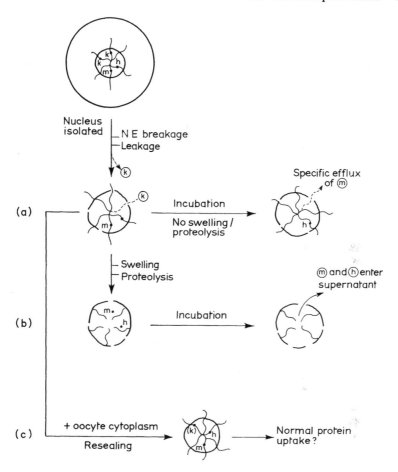

Figure 2.9 Use of isolated nuclei to study transport processes. Top: the nucleus *in situ* has fibrillar connections into the cytoplasm and contains, among other things, karyophilic proteins (k), mature mRNA (m) and HnRNA (h). On isolation, the nucleus loses many of the karyophilic proteins because of nuclear envelope breakage. Incubation in aqueous media allows specific efflux of mature mRNA and retention of HnRNA if swelling is prevented and proteinase and RNAase activities are inhibited (a), but if these conditions are not met then both mRNA and HnRNA detach from their intranuclear binding sites and leak into the supernatant (b). On addition of oocyte cytoplasm to the isolated nucleus, the envelope reseals and apparently normal uptake of karyophilic proteins can be demonstrated (c).

returning the nuclei to an aqueous environment – whereupon the problems that have been evaded by the use of the organic solvent re-emerge.

In short, experiments in which isolated nuclei are incubated in aqueous buffers, and the distribution of macromolecules between nucleus and medium is subsequently examined, can scarcely be expected to provide interpretable results. Many such experiments have in fact been performed, and they are largely ignored by most workers in the field, who prefer the careful and informative *in situ* studies described in the first section of this chapter.

Nevertheless, against all the odds (and against the weight of opinion in the field), there seem to be two kinds of situations in which the results of transport experiments with isolated nuclei reflect genuine physiological events. The first situation concerns some, but by no means all, experiments on mRNA export from isolated nuclei. The second situation concerns experiments in which the isolated nuclei are suspended not in an ordinary aqueous buffer, but in an amphibian oocyte cytoplasm preparation; in this medium the broken NE apparently reseals and something closely akin to *in situ* transport properties are restored, so that aspects of specific protein import can be studied. In both situations, the advantages over *in situ* studies are clear: the system is much simpler than the whole cell, and parameters such as temperature, ATP concentration, specific ion activities and so on can be fixed by the experimenter at any value he or she desires. In so far as the results pertain to physiological transport processes, they can therefore lead to far more detailed characterization than can be achieved with almost any *in situ* study; in particular, values for dissociation constants of permeant binding and 'K_ms' for 'transport' are much more reliable. Moreover, the techniques involved are usually simpler, less expensive, and quicker to execute than those associated with *in situ* studies.

The development of *in vitro* methods for studying mRNA transport has been reviewed in detail (Agutter, 1988); the salient points will now be summarized. The history of such studies with isolated nuclei dates back to the work of Schneider (1959), about a year before the actual existence of mRNA was definitively demonstrated, but widespread use of such *in vitro* procedures began around 1970. Throughout the 1970s, protagonists of these methods habitually based claims about the physiological relevance of their findings on very general properties of the RNA in their postnuclear supernatants after incubation and centrifugation: sedimentation rate, poly(A) content, kinetics of hybridization with DNA, capacity to support *in vitro* translation, and so on. None of these data answered questions about the general integrity of any mature mRNAs in the supernatants, or about the extent of contamination with incompletely processed mRNA precursors. In short, it was not demonstrated that anything akin to normal nuclear restriction was maintained in such experiments. Indeed, some of the many nuclear incubation media resulted more or less in complete lysis of the nuclei when ATP or other chelating agents, or exogenous polyribonucleotides, were added. This situation led to an entrenchment of the repeated criticisms of *in vitro* studies.

This entrenchment has persisted, although more substantial support for at least some *in vitro* methods has been provided since 1980. Studies in several laboratories have shown that something approximating to normal nuclear restriction of mRNA and its precursors is maintained during such experiments provided that two criteria are met: the nuclei are not allowed to expand; and proteinase and RNAase activities in the preparations are inhibited as far as possible. Subject to these conditions, the following salient pieces of evidence for normal restriction have been obtained for hepatocyte, hepatoma, myeloma and other kinds of mammalian nuclei.

1. Splicing intermediates that can be identified unambiguously on Northern blots, such as the precursors of immunoglobulin heavy-chain messengers, are absent from the supernatants but remain detectable in the nuclear pellets after incubation.
2. The supernatant RNA from hepatocyte nuclei forms the three abundance classes characteristic of cytoplasmic messengers, and hybridization with cDNA probes against nucleus-restricted sequences shows significant contamination only in the lowest abundance class.
3. The core A, B and C group proteins characteristic of HnRNP particles are virtually absent from the supernatants.
4. Contamination of the supernatant RNA with cytoplasmic messengers desorbed from the nuclear surface (rather than exported from the nuclear interior) does not exceed 15% of the total.
5. Pulse-labelling studies show a delay in the export of labelled mRNA compatible with the time taken to complete post-transcriptional processing.

These (and other, supporting) lines of evidence make claims about the physiological significance of some *in vitro* experiments perfectly credible. On the face of it, this conclusion is surprising in view of the non-physiological nature of isolated nuclei. In fact, it is exactly what the solid-state model of mRNA transport would predict, and it constitutes one of the strongest arguments in favour of this model (Figure 1.3(c)).

However, any experiment of this kind has to be examined critically. First, it needs to be ensured that the basic requirements – maintenance of normal nuclear size and inhibition of degradative enzymes – have been met. Second, data pertaining to low-abundance messengers remain very dubious, because contamination of these with nuclear precursors does seem to be significant. Third, any extrapolation to other kinds of cell nuclei is inherently unwise: detailed characterization of the method needs to be carried out with each new source of nuclei examined – the fact that such-and-such a medium sustains near-normal restriction in liver nuclei does not mean that the same medium will do so in brain nuclei, even from the same animal. Fourth, and most important, it seems inappropriate to describe the phenomenon observed in an *in vitro* experiment as 'nucleocytoplasmic mRNA transport', partly because the isolated nucleus is a mere shadow of its *in vivo* self, and partly because the medium into which the mRNA is being exported does not by any stretch of the imagination correspond to 'cytoplasm'. In terms of the nomenclature introduced in Chapter 1, the results of such an experiment will at best reflect the release and translocation stages of transport; the cytoskeletal binding stage is absent. It is generally better to describe the phenomenon by the phrase 'mRNA efflux'.

The use of amphibian oocyte cytoplasm to reseal isolated nuclei was pioneered by Newmeyer *et al.* (1986) (Figure 2.10). The justification for the technique rests primarily on two findings: the observation by Forbes *et al.* (1983) that DNA (irrespective of source) injected into oocyte cytoplasm immediately becomes surrounded by ultrastructurally normal NE, presumably formed from cytoplasmic precursor pools in response to the stimulus provided by the DNA itself; and the reconstitution of membrane-depleted *Xenopus* spermatozoa by yolk-depleted *Rana* cytoplasm (Lohka and Masui, 1983). Newmeyer and his colleagues have examined the properties of their isolated resealed nuclei with great care, and have demonstrated close comparability with homologous nuclei *in situ* in respect of the specificity and kinetics of protein import and of general morphological characteristics. The results will be discussed in detail in Chapter 4. In all important respects so far described in the literature, Newmeyer's nuclei

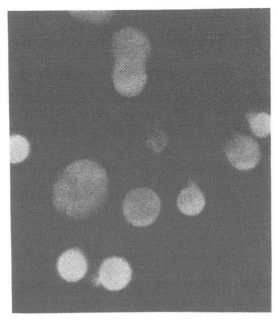

Figure 2.10 An isolated hepatocyte nucleus resealed in *Xenopus* oocyte cytoplasm. Photograph kindly supplied by Dr D. D. Newmayer and reproduced by permission of the Editorial Board of *J. Cell Biol.*

behave physiologically. This distinguishes them from all systems in which isolated nuclei are resuspended in aqueous buffers, where behaviour is more or less physiological only in respect of mRNA (and possibly of ribosome) efflux, and movements of soluble proteins, tRNA and other permeants simply cannot be studied. The fact that other workers in the field have not adopted Newmeyer's simple and reliable procedure might partly be attributed to its novelty, but might also reflect the deep mistrust of all *in vitro* methods that remain widespread in the field.

Nevertheless, two caveats need to be borne in mind in interpreting results from such experiments. First, Newmeyer's nuclei still suffer from many of the defects common to all isolated nuclei: loss of soluble protein components, a measure of proteolytic and other degradation, and a measure of cross-linking. These factors might prove important in some experiments. Second, oocyte cytoplasm is an ill-characterized medium, and in some experiments it is not clear that all relevant parameters are properly known. For example, if a putative transport inhibitor is added, what is its effective concentration? How much of it has become bound to the cytoplasmic proteins? If ATP is added, how rapidly is it hydrolysed by components of the cytoplasm? And so on.

2.2.2 Resealed nuclear envelope vesicles

Even in the most reliable *in vitro* experiments with isolated nuclei, one of the problems endemic among *in situ* methods persists: the problem of knowing when a result can be attributed to intranuclear binding or release events, and when it must be attributed to translocation events at the NE (specifically the PCs). Ideally, this problem could be solved by the use of whole NEs, free of intranuclear material, resealed to form nucleus-sized vesicles which have *in vivo* permeability properties. Results obtained with such a system would necessarily reflect translocation events, not intranuclear events.

An approximation to this ideal was first described by Fasold's laboratory (Kondor-Koch et al., 1982) and a more detailed characterization of the preparation was published recently by the same group (Riedel et al., 1987). Briefly, the procedure is as follows (Figure 2.11). Only one of the numerous methods for isolating NEs (section 2.3.1 gives a further discussion) seems to provide envelopes suitable for resealing: the method of Bornens and Courvalin (1978), in which the chromatin in the isolated nuclei is solubilized by the addition of heparin in the presence of phosphate. Fasold and his colleagues found that the NEs recovered from the lysate by centrifugation could be resealed by the addition of calcium ions to a concentration of around 3 mM; and furthermore, that macromolecules such as proteins and poly-ribonucleotides, and even complexes such as HnRNP particles and ribosomal subunits, could be entrapped in the vesicles if they were added prior to resealing.

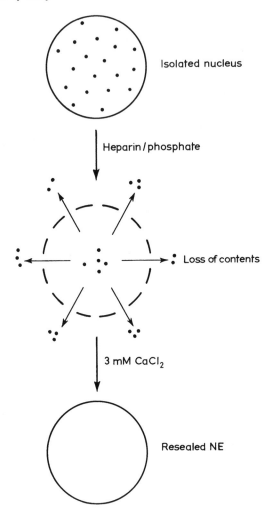

Figure 2.11   Preparation of resealed nuclear envelope vesicles. On the addition of heparin in phosphate buffer to a final heparin:DNA ratio of 3:2 the nuclei swell dramatically and most of their internal contents are solubilized via breaks in the envelope. After gentle washing and recovery, the almost-empty nuclear 'ghosts' are resealed with approximately 80% efficiency by addition of $CaCl_2$.

There is no doubt that the resealing process is fairly efficient; the vesicles exclude at least some materials that cannot penetrate the NE *in situ*, and they take up histones and high mobility group proteins, but not immunoglobulins, myoglobin or cytochrome c from the incubation medium. They export mRNA in a manner that seems to be dependent on ATP hydrolysis and susceptible to specific inhibitors, in agreement with results obtained from the best-characterized isolated nuclei systems. Vesicles that are deliberately made leaky, for instance by freezing and thawing or by homogenization, lose these interesting permeability properties. For these reasons, the resealed NE vesicle technique seems likely to provide another useful approach to the study of nucleocytoplasmic transport processes.

However, as for all techniques, there are important criticisms. Although the majority of vesicles are resealed in a good preparation, a substantial minority (typically 15–20%) remain leaky. This means that any set of results is likely to have a high background, and despite the use of careful controls only quite marked effects can be measured. Moreover, it would be naïve to suppose that the vesicles were completely free of intranuclear contents. On average, they retain about 1.5% of the total nuclear DNA, and this fact alone is sufficient to show that effects observed in vesicle experiments might still involve some intranuclear binding as well as translocation events. Finally, the biophysical studies performed by Lang and Peters (1984) using the laser photolysis technique showed that a measure of PC damage is inevitable when nuclei are isolated and more marked when envelopes are isolated. This effect is indicated by an apparent increase in the mean patent pore radius of some 5–10%, which has profound implications for the permeability properties of the NE. Peters has commented that the complete removal of one single PC from a hepatocyte NE, leaving a PC-sized hole in the envelope, would more or less double the rate of permeation of small proteins. Given that the NEs used to make resealed vesicles must have suffered some such damage, which might or might not have been partly reversed by the resealing process, even the encouraging results obtained so far have to be interpreted with some caution.

As with Newmeyer's method, the resealed vesicle technique is novel and has not, despite its simplicity and reproducibility, been used in many other laboratories so far. This situation makes any attempt at evaluation somewhat premature, and although encouraging and interesting results have been obtained with the technique, caution is undoubtedly necessary.

2.3 STUDIES ON NUCLEAR ENVELOPES AND OTHER SUBNUCLEAR STRUCTURES

2.3.1 Nuclear envelope isolation

Methods for isolating nuclei in bulk from such tissues as mammalian liver developed during the 1950s and matured in the 1960s with the emergence of procedures such as that of Blobel and Potter (1966), which remains the most widely used. Methods for isolating NEs emerged in the wake of this development; the first ones (apart from microdissection techniques applicable to amphibian oocytes, which were described somewhat earlier) were published around 1970. Numerous NE isolation methods have been described since then, each with its own particular advantages and disadvantages. Broadly, they fall into two groups: low ionic strength methods and high ionic strength methods.

Most low ionic strength methods rely on disruption of the chromatin by DNAase treatment, followed by repeated washing of the envelopes. An exception is the heparin procedure of Bornens and Courvalin (1978) mentioned earlier. By far the most widely-used low ionic strength approach is the one described by Kay *et al.* (1972) (Figure 2.12a), in which the isolated nuclei are subjected to DNAase I treatment first at pH 8.5, then at pH 7.5. The point of the initial high pH is to destabilize the chromatin and thus to increase the efficiency of action of the DNAase. Morphologically well-preserved NEs with low intranuclear contamination are obtained, but the serine proteinases that always abound in liver nuclei might play some part in the removal of the nuclear contents. Certainly, some workers report difficulties in getting the Kay procedure to work when the buffers are supplemented with PMSF, but others have no such difficulties; the reasons for this discrepancy are not clear. Another problem of the Kay method for some studies is the use of magnesium ions, which are indispensible co-factors for DNAase I; Mg^{2+} might alter the PC morphology (Maul, 1977) and artefactually deposit intranuclear components on the NEs. This point was the basis for a more recently-published low ionic strength method (Krachmarov *et al.*, 1986), which avoids this problem by utilizing DNAase II in the presence of chelating agents. Whole nuclear 'ghosts' are obtained, and again the intranuclear contamination seems to be low, but as yet this method has not been widely used.

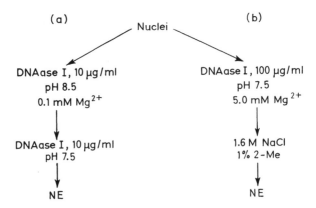

Figure 2.12 The two nuclear envelope isolation procedures described in the text. (a) the Kay *et al.* method; (b) the Kaufmann *et al.* method.

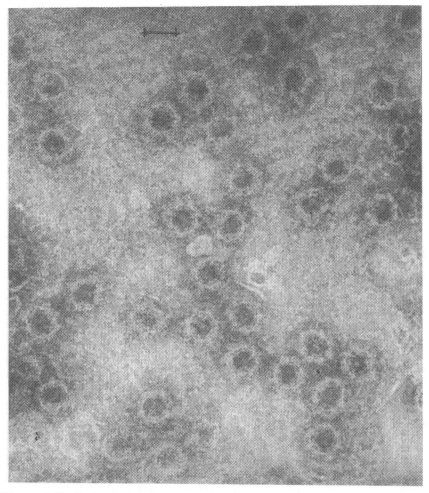

Figure 2.13 Isolated nuclear envelopes revealed by electron microscopy. The specimen was negatively stained with 2% ammonium molybdate. Scale bar = 100 nm. Photograph by the author.

Among the many high ionic strength procedures, the one described by Kaufmann *et al.* (1983) (Figure 2.12b) seems the most satisfactory in terms of shortness of operation time, integrity of the NEs, and protection of the endogenous proteins against degradation and artefactual cross-linking. In essence, the technique involves disruption of the chromatin with DNAase I and β-mercaptoethanol at neutral pH, followed by solubilization of the nuclear contents in 1.6 M NaCl. Putative NE components that seem to be lost in low ionic strength procedures seem to be preserved by this approach,

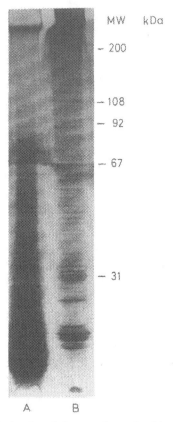

Figure 2.14 Sodium dodecyl sulphate polyacrylamide gel electropherograms of nuclear envelopes isolated from rat liver by the methods of (a) Kay *et al.* and (b) Kaufmann *et al.* Electrophoresis was performed by the Laemmli method using a 12% separating gel, and the proteins were stained with Coomassie blue. Note the predominance of the lamins (60–70 kDa range: see Chapter 3). The preparations and photographs are by Dr S. J. M. Aitken and the author.

but the problems of using Mg^{2+} and possible subtle damage to the NE consequent on exposure to the high salt concentration cannot be avoided.

Judgements of quality for all isolated NE preparations (Figure 2.13) depend on demonstrations of morphological integrity and of low cytoplasmic and intranuclear contamination. These criteria are not rigorous. However, it is reassuring to note that biochemical analyses of NEs isolated by these various procedures show good general agreement, at least in terms of the lipid components and of the polypeptide composition as revealed by SDS-polyacrylamide gel electrophoresis (SDS-PAGE) (Figure 2.14).

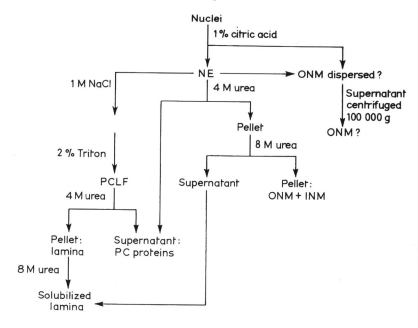

Figure 2.15 Summary of the main procedures employed to date for subfractionation of the nuclear envelope. Citric acid (1%) disperses the ONM from both isolated nuclei and NE, leading to putative ONM preparations. Sequential NaCl and Triton extration removes both membranes and leaves the PCLF. Extraction of either isolated NE or PCLF with 4 M urea solubilizes the PC components, apparently specifically. When the PCLF is treated in this way, the isolated lamina survives and can be solubilized with 8 M urea. Extraction of the NE with 8 M urea removes the lamina as well as the PCs, and leaves the membranes (INM and ONM).

2.3.2 Subfractionation of the nuclear envelope (Figure 2.15)

Because so much of the interest in NE structure and function centres around the PCs, attempts to isolate PCs free of other envelope components date back more or less to the time that isolated NE preparations became available. This ambition has never been achieved, and it is now widely accepted that isolation of PCs as intact, morphologically identifiable entities may be impossible in principle. Just what this inference implies about the nature of the PCs will be discussed in Chapter 3.

Blobel's laboratory was in the forefront of attempts to isolate the PC in the early 1970s, and their efforts led to a procedure for isolating PCs attached to the lamina but essentially free of both membranes (the ONM and INM). This preparation, the pore-complex-lamina fraction (PCLF) (Figure 2.16),

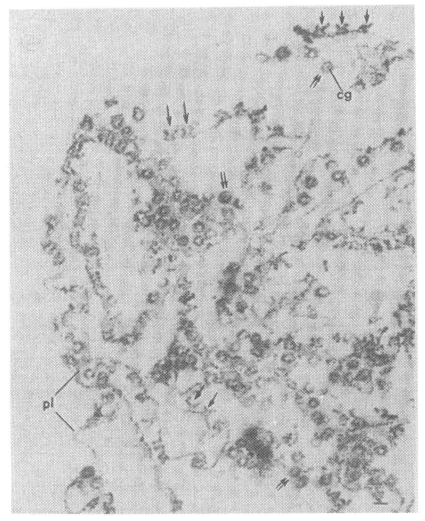

Figure 2.16 The isolated PCLF from rat liver. Photograph kindly supplied by Dr G. Blobel and reproduced by permission of the Editorial Board of *Proc. Natl. Acad. Sci. USA*. Bar = 100 nm.

was first made from rat liver NEs isolated by the Kay procedure (Aaronson and Blobel, 1975), and it has since been applied to many different cell types. In essence, the NEs are treated first with a high ionic strength solution (1–2 M NaCl) and then with a high concentration of a non-ionic detergent, routinely 2% Triton X-100. The high detergent concentration is necessary to solubilize the INM phospholipid (and other) components. The PCLF fraction is morphologically well defined and has been immensely valuable in furthering our knowledge of NE biochemistry.

Lamina free of PCs can be made from the PCLF by extraction with 4 M urea (Maul and Baglia, 1983). It seems that the distinctive components of the PC are solubilized by the urea; this treatment also removes PCs from isolated NEs. Two applications of the urea treatment have been: (a) as aids in the biochemical characterization of PC components, and (b) to provide controls in *in vitro* permeability studies using resealed NE vesicles (Figure 2.17) (removal of the PCs destroys the permeability barrier: cf. Lang and Peters,

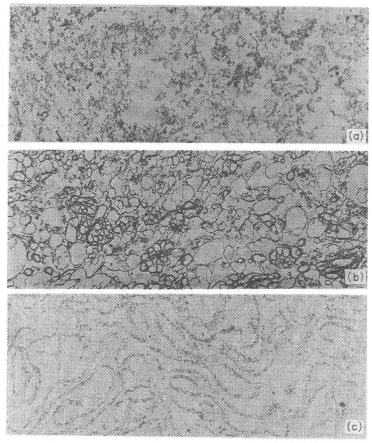

Figure 2.17 Nuclear envelopes were isolated from the germinal vesicles of *Spisula* oocytes and either the NE itself or the PCLF obtained from it was extracted with 4 M urea. Note the disappearance of the PCs but the persistence of the lamina. Magnification: (a) × 15 000, (b) × 15 000, (c) × 19 000. Photographs kindly supplied by Dr G. G. Maul and reproduced by permission of the Editorial Board of *Exp. Cell Res.*

1984). The lamina itself can be solubilized at higher urea concentrations (around 8 M) under reducing conditions.

There is no consensus about methods for obtaining distinct ONM and INM preparations, but an approach that has been used exploits the long-established fact that isolated nuclei treated with 1% citric acid appear to lose their ONM (Gurr *et al.*, 1963). Undoubtedly the citric acid disperses membrane material, just as it disperses ER membranes. After treatment, the nuclear surface appears to be a single-membrane structure on electron micrographs of thin sections, and the nuclei have lost roughly half their phospholipid. The dispersed membrane can be recovered from the post-nuclear supernatant by centrifugation at 100 000 g. When isolated NEs are extracted with 1% citric acid, the results are similar: the ONM, but not the INM, appears to be dispersed. Doubts about the value of the procedure stem from the fact that there is no established marker for either ONM or INM, and the extent of cross-contamination therefore cannot be estimated. Moreover, citric acid might well solubilize membrane components or artefactually precipitate contaminants of non-membraneous origin on to the membranes. Nevertheless, the 'ONM' isolated in this way is biophysically different from the 'INM' (its phospholipid has a much lower order parameter as measured by electron paramagnetic resonance using a stearic acid-derived probe; cf. Agutter and Suckling, 1982). It is also, according to a personal communication from Mel Schindler, biochemically different: it contains none of the inositol phospholipids of the NE.

2.3.3 The nuclear matrix

Discussion of the whole controversy about the nuclear matrix would be inappropriate here; only some general methodological points will be considered – though these are, in fact, fundamental to the controversy. The nub of the issue is that the phrase 'nuclear matrix' has been used to describe at least two kinds of entity, one *in vivo* and one *in vitro*, and the relationship between the two is unclear and at best indirect. To avoid confusion, the term **nucleoskeleton** (NS) will be used for the *in vivo* entity putatively identified by electron microscopy, and the term nuclear matrix will be reserved for certain kinds of controversial *in vitro* preparations; but this is not common practice, and readers should be aware that these and other names are not used consistently in the literature. In the context of this book, the NS/nuclear matrix issue is relevant to nucleocytoplasmic transport because of the potential role of such a structure in the intranuclear binding of RNAs and karyophilic proteins (cf. the discussion of Feldherr's work in section 2.1.3). Broadly speaking, cell biologists tend to accept the reality of the NS/matrix, while most molecular biologists see no need to postulate any such structure.

Although intranuclear structures analogous to the cytoskeleton had been mentioned occasionally in the literature since the 1940s, the issue only became controversial when Berezney and Coffey (1974) described a method for isolating the nuclear matrix from rat liver. The cell type used and the procedures employed were exactly the same as those used by Aaronson and Blobel (1975), almost exactly contemporaneously, to isolate the PCLF. Yet while Aaronson and Blobel obtained a structure that contained only the non-membraneous parts of the nuclear periphery, Berezney and Coffey obtained something quite different (Figure 2.18). They interpreted electron micrographs of sections of their preparations in terms of a broad, anastomosing

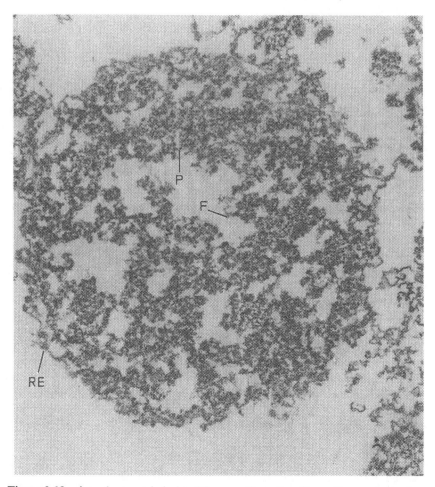

Figure 2.18 A nuclear matrix isolated from rat liver: F = fibres, P = pore complex, RF = residual envelope. Magnification: × 14 000. Photograph kindly supplied by Dr R. Berezney and reproduced by permission of the Editorial Board of *J. Cell Biol.*

network of thick, amorphous fibres running through the whole nuclear interior and including a nucleolar residue as well as the PCLF. Subsequent studies such as those of Kaufmann *et al.* (1981) helped to explain this conflict. Briefly, it seems that the stability of the apparent intranuclear structure depends on: (a) oxidative cross-linking of the component proteins – the effect can be achieved by adding sulphydryl oxidizing reagents – and (b) treatment with RNAases after, rather than before, the high salt extraction. If the nuclear RNA is attacked early in the process and oxidation is not extensive (e.g. the operation time is short), then the PCLF rather than a nuclear matrix is obtained. This implies that the nuclear matrix is simply an artefact of a long *in vitro* procedure, though this is difficult to reconcile with the fact that the nuclear matrix is associated with a wide range of specific nuclear functions such as the initiation of DNA replication (Berezney, 1979). The importance of the timing of RNAase treatment might imply that the nuclear matrix is an aggregate of HnRNP particles, which are certainly prone to form artefactual fibrillar complexes in high salt media (see e.g. Arenstorf *et al.*, 1984), though the protein compositions of both HnRNPs and nuclear matrices are too complex and ill-characterized for rigorous comparison to be possible.

Nuclear matrices isolated without the use of RNAase and with the addition of serine proteinase inhibitors contain most or all of the nuclear RNA (Long *et al.*, 1979), which lends credence to this implication. If the matrix is a real *in situ* structure to which RNPs are bound, it is difficult to explain why individual HnRNP particles with no associated fibrillar structure can be visualized, e.g. on Miller spreads, in which nuclei are lysed in water directly on to an electron microscope specimen grid. These considerations had led by the early 1980s to a widespread conviction that the nuclear matrix was a methodological artefact, and therefore (this is a logically unsound inference) that nuclei *in vivo* have no structural framework analogous to the cytoskeleton. Nevertheless, the fact that nuclear matrix preparations are associated with such a wide range of nuclear functions has led many people to continue using them; such workers are not always explicit about the fact that their 'nuclear matrices' are at best only operationally defined.

Evidence for the *in vivo* NS postdated this controversy, and the demonstration was greeted sceptically (or ignored) because by that time the conviction that all 'intranuclear fibrillar systems' are artefactual had become entrenched. However, at least two groups of studies have provided good support for the belief that the NS is real. First, when liver cells are treated with the transcription inhibitor α-amanitin, the chromatin contracts transiently, mainly on to the nuclear periphery (Figure 2.19). On thin sections, the spaces vacated by the euchromatin were occupied by a dense, anastomosing, reticular meshwork of very fine and homogeneous fibrils (Brasch, 1982), quite different in appearance from the isolated nuclear matrix described by Berezney and Coffey. Second, a similar reticular intranuclear meshwork has been repeatedly visualized by electron microscopy in resinless sections of untreated cells (e.g. Fey *et al.*, 1984). This structure, called the 'nucleo-skeleton' here, was dubbed 'nuclear matrix' by its discoverers, who thereby prejudiced the acceptability of their findings to an audience which was already conditioned to be sceptical about nuclear matrices. The resinless

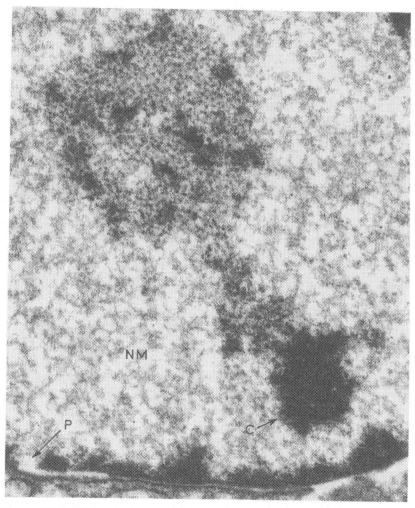

Figure 2.19   Electron micrograph of a nucleus in a cross-section of α-amanitin treated liver. The NS is clearly visible in the regions vacated by the highly condensed chromatin: C = chromatin, NM = nuclear matrix, P = pore complex. Photograph kindly supplied by Dr K. R. Brasch and reproduced by permission of the Editorial Board of *Exp. Cell Res.*

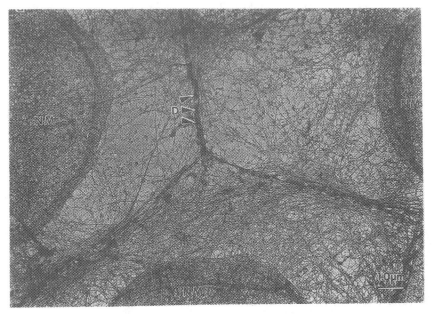

Figure 2.20 The 'nuclear matrix-intermediate filament network' visualized in resinless sections: NM = nuclear matrix. Photograph kindly supplied by Dr S. Penman and reproduced by permission of the Editorial Board of *J. Cell Biol.*

section studies (Figure 2.20) showed that the NS was continuous through the PCs with the cytoskeleton, an observation that has interesting implications for the solid-state model of nucleocytoplasmic transport.

Although these results support belief in the NS, they are not completely unequivocal. The α-amanitin studies reveal a structure in liver cells that have been subjected to considerable toxic damage; and although the resinless section micrographs obtained by Penman and his colleagues are of superb quality, this technique is fraught with difficulties of interpretation. There are therefore rational grounds for continuing scepticism. For the purposes of this book, the reality of the NS will be assumed but this assumption remains controversial.

If the NS is real, why is it not seen on Miller spreads (Figure 2.21), or in NEs made by (for example) the Kay procedure? Why is there no evidence for it in nuclei treated with heparin (Bornens and Courvalin, 1978) or DNAase II and chelating agents (Krachmarov *et al.*, 1986) at low ionic strength? This is rather like asking why cytoskeletal structures are not apparent in conventional tissue homogenates. A possible answer lies in the evidence that the NS contains short actin filaments (Nakayasu and Ueda, 1984) and that its structural integrity is dependent on these. F-actin dissociates at low ionic

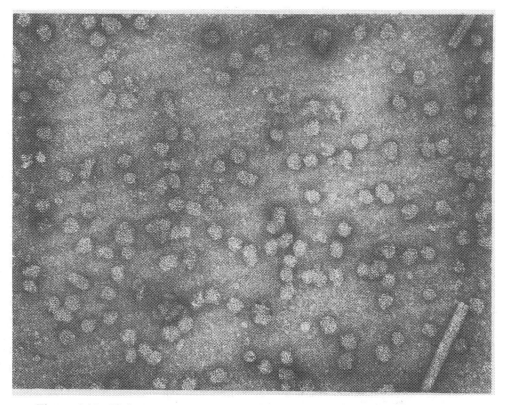

Figure 2.21 HnRNP particles from HeLa cells after sucrose density gradient fractionation. Miller spreads of nuclei normally show these 'chain-of-beads' structures but no evidence of a NS or similar structure. Photograph kindly supplied by Dr W. M. LeStourgeon and reproduced by permission of the Editorial Board of *J. Cell Biol.* Bars = 50 nm.

strength, which could account for the non-appearance of the NS in such circumstances. Moreover, the NS is no doubt a very delicate structure, and the disruption or removal of the bulk of the chromatin – a procedure that is in effect universal in subnuclear fractionation – would affect it in much the same way that a tornado affects a chickenwire fence.

Since the first accounts of the *in situ* NS were published, two procedures for isolating an ultrastructurally similar entity from nuclei have been described (Figure 2.22(a), (b), (c)) (Comerford *et al.*, 1986; Fey *et al.*, 1986). Reassuringly, this preparation is disrupted by actin-dissociating agents such as cytochalasin B, and individual HnRNP particles can be recovered from it. Less reassuringly, monoclonal antibodies that seem to decorate the NS *in situ* (Chaly *et al.*, 1984) have not yet been shown to react with any of its specific

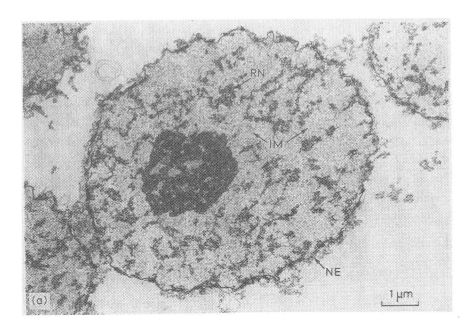

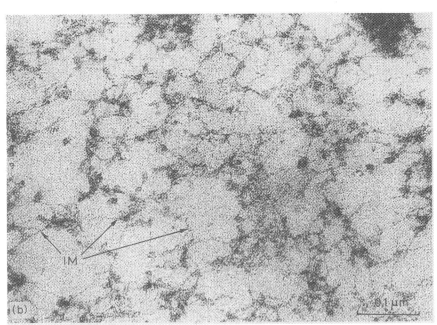

Figure 2.22 A 'nuclear reticulum' isolated from rat liver by the Comerford *et al.*
procedure. (a) and (b) show the intact structure at low and high magnification;

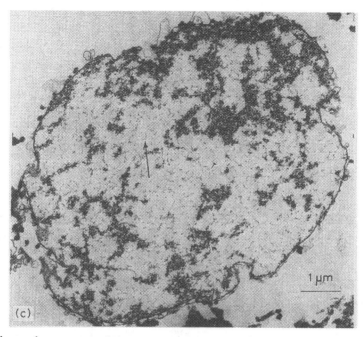

(c) shows the structure (low magnification) after treatment with cytochalasin B: IM = internal matrix, NE = nuclear envelope; RN = residual nucleolus. Photograph by Ms S. A. Comerford, reproduced by permission of Dr A. S. MacGillivray and Butterworth Ltd, London.

polypeptide components. Its composition seems to be very complex. If a nucleoskeletal element is analogous to a cytoskeletal element, a small number (1–4) of core fibril components would be expected; the evidence that DNA topoisomerase II and actin are dominant NS components (Chapter 3) does not amount to a completely satisfactory characterization. Of course, there are at least three kinds of cytoskeletal elements; so why should there be only one kind of nucleoskeletal element? Might the apparent complexity and ill-characterized composition of the isolated preparation result from the inherent heterogeneity of the NS? Possibly; but it would be unwise in our present state of ignorance to claim that these isolated structures are 'identical' with the *in situ* NS, despite the encouraging structural and other properties. To maintain clear distinctions in the field, we need a name that distinguishes them from both the NS and the nuclear matrix, and elsewhere the name **nuclear reticulum** has been proposed for this purpose (Agutter, 1988).

Readers who are interested in further methodological details about the NE, NS and other nuclear components relevant to nucleocytoplasmic transport (among other fields of research) should consult the compilation by Birnie and

MacGillivray (1986), several chapters of which are directly relevant to the issues discussed in this book. Schröder *et al.* (1989) have recently published a detailed methodological survey of studies directly related to RNA transport.

2.4 OVERVIEW

In this chapter the methods for studying nucleocytoplasmic transport processes and the nuclear structures relevant to these processes have been surveyed. Nucleocytoplasmic transport has been studied by both *in situ* and *in vitro* methods. *In situ* methods have been most widely used and are fundamentally more reliable, because the NE remains intact (unless it has been deliberately disrupted), the PCs ultrastructurally normal, and the environment of the NE relatively unperturbed on both the nucleoplasmic and cytoplasmic sides. The main limitation of such methods lies in the great complexity of the cell's internal environment and the experimenter's inability to know exactly – much less to vary as he or she might wish – the biochemical parameters relevant to detailed understanding of the transport machinery. These methods have been used to measure the functional pore radius of the PCs, and (in combination with genetic engineering and monoclonal antibody techniques) to identify structural features in macromolecules that make them, for example, transportable or karyophilic. Permeabilization of cells to agents that interfere with the transport machinery, e.g. antibodies against NE components, or microinjection of such agents, has also helped to gain some insight into transport mechanisms. But overall, studies of this kind are capable of providing only qualitative rather than quantitative results.

In vitro methods are limited because of the damage inevitably suffered by nuclei during isolation. Three kinds of methods have been reviewed. Isolated nuclei suspended in aqueous buffers can be used to elucidate mRNA transport – but not, for example protein or tRNA transport – provided nuclear expansion and degradative processes are prevented. Resealed NE vesicles are a promising tool for some purposes, though a rather high level of background leakage has to be accepted. Finally, isolated nuclei resealed by suspension in amphibian oocyte cytoplasm appear to behave physiologically, and may afford advantages over *in situ* systems in terms of the possibilities for controlling relevant biochemical parameters.

The structures discussed in this chapter have been the NE itself and the NS/nuclear matrix. Among the bewildering array of NE isolation procedures, the low ionic strength Kay *et al.* method and the high ionic strength Kaufmann *et al.* method are the most widely-used, mainly because of the morphological integrity of the NEs produced and the low levels of identifiable contaminants; and the heparin method of Bornens and Courvalin

(1978) is valuable because the NEs made by it can be resealed. Some NE subcomponents can be isolated: the lamina, the pore-complex-lamina fraction and possibly the ONM. PCs and the total membrane (ONM + INM) can be solubilized, apparently specifically.

The existence of an *in vivo* NS is reasonably well established and will be assumed henceforth. This structure might *inter alia* be important in binding proteins and nucleic acids within the nucleus and therefore be relevant to nucleocytoplasmic transport. The isolated nuclear matrix preparation seems to be only indirectly related to the NS, but it remains widely used because of its association with specific nuclear functions. Isolated preparations (nuclear reticula) morphologically similar to the NS have been obtained. They may be of greater experimental value than 'classic' nuclear matrix preparations, but as yet they have not been adequately characterized.

Methods for isolating subcellular structures are usually preconditions for biochemical analysis. A good deal is known about the biochemistry of the NE, and there is some provisional information about the molecular organization of the NS. The next chapter will be devoted to examining those parts of our knowledge of the NE and the NS/nuclear matrix that help to throw light on the nucleocytoplasmic transport of macromolecules.

3 Intracellular structures in nucleocytoplasmic transport

Most of this chapter will be devoted to the nuclear envelope and its components. The little information we have about the role of the NS and cytoskeleton in nucleocytoplasmic transport will be considered briefly in the final section.

3.1 NUCLEAR ENVELOPE ULTRASTRUCTURE

The NE, with its ubiquitous and morphologically complex PCs, is the key structure in nucleocytoplasmic transport. In most eukaryotic cells it is disassembled and reassembled at mitosis. For these and other reasons, the NE has interested electron microscopists since 1949 and continues to do so. Our understanding of NE ultrastructure has developed with and in consequence of the development of the electron microscope *per se* and of techniques for specimen preparation, staining and examination. This coevolution of technique and concept, method and understanding, can be seen by studying the series of major reviews devoted to the NE; those by Gall (1964), Franke (1974), Maul (1977) and Franke *et al*. (1981) are particularly authoritative – these reviewers have been among the most important contributors to the study of the NE and its morphology. Important individual papers include the comparative study of specimen preparation techniques by Abelson and Smith (1970), the account by Unwin and Milligan (1982) which includes possibly the most highly resolved images of the PC to date (Figure 3.1), and a series of articles by Maul and his colleagues discussing the factors that affect numbers of PCs on the nucleus and identifying possible sources of error in interpreting micrographs (see Maul, 1977, 1982 for reviews).

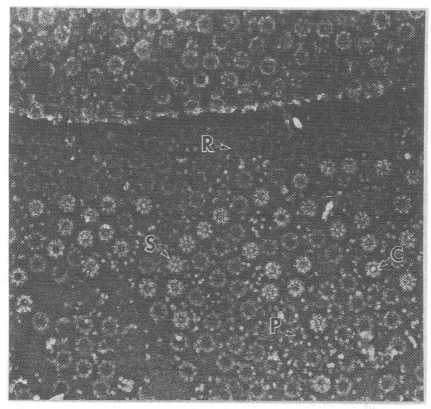

Figure 3.1 Electron micrograph (uranyl acetate staining) of nuclear pore complexes from Unwin and Milligan (1982): C = central plug, P = pore lumen, S = spokes. Magnification = × 60 000. Photograph kindly supplied by Dr P. N. T. Unwin and reproduced by permission of the Editorial Board of *J. Cell Biol.*

The main electron microscopic techniques that have been deployed in studies of the NE are: (a) positive staining of dry whole mounts; (b) negative staining; (c) thin sectioning after osmium tetroxide fixation; (d) freeze-etching; and (e) scanning electron microscopy of whole desiccated nuclei. Certain other techniques, e.g. $KMnO_4$ fixation, have been discontinued because of specimen damage; elaborations of, for example negative staining studies such as the Markham rotation procedure have been used in the analysis of structural symmetries. The breadth of this range of techniques largely accounts for disagreements of detail between ultrastructural accounts of the NE; what is more impressive is the extent of general agreement between users of the various methods.

3.1.1 The pore complex

Despite the proliferation of structural models (Arçhega and Bahr, 1985; see also Fig. 1.4), certain general features of the PC are now almost universally agreed. These are: the size of the structure (generally 90–100 nm in external diameter); the overall configuration – the PC is a cylinder with its axis orthogonal to the membrane plane, the wall of the cylinder, the **annulus** (Figure 3.1), being up to 25 nm thick; the octagonal symmetry apparent when the annulus is seen 'end-on', i.e. from the cytoplasmic or the nuclear surface; the fact that the annulus extends beyond the planes of the membranes on both cytoplasmic and nucleoplasmic surfaces; and the fact that the annulus overlaps the edges of the ONM and INM, which seem to be continuous with each other at this point. Each of the eight subunits of the annulus has a bulky component at its outer extremity on the cytoplasmic surface (Unwin and Milligan, 1982). Figure 3.2 summarizes these features in a 'minimal' structural model of the PC.

Other features are more controversial. The fine structure of the annular

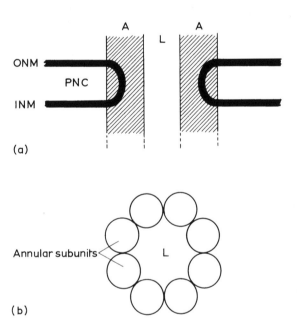

Figure 3.2 'Minimal' schematic representation of a nuclear pore complex (PC) in (a) vertical section and (b) surface view. The outer and inner nuclear membranes (ONM and INM) are continuous at a point overlapped by the annulus (A) of the PC. The lumen (L) is portrayed as empty, though many authorities would include a central granule/tubule attached either to the annular subunits or to the membranes by a diaphragm or fibrils (struts). PNC = perinuclear cisterna.

subunits has been variously represented. Abelson and Smith (1970), for example, portrayed the subunit as a 'minitubule' with fibrillar walls extending from nucleus to cytoplasm; Franke (1974) declared it to be a ribosomal particle, and others have similarly questioned its continuity through the NE plane; Maul (1982) gave evidence suggesting that its apparent thickness and particulate appearance were artefacts caused by Mg^{2+} ions – the 'annular subunit' is really a '**traverse fibril**' crossing the NE which rolls up into a 'particle' in the presence of Mg^{2+} (Fig. 3.3). Most of these ideas have been met with general scepticism, though Maul's view merits serious consideration in the light of evidence for NS-cytoskeleton continuity

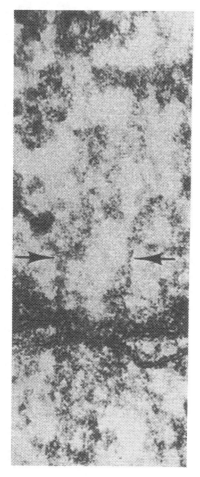

Figure 3.3 Electron micrograph showing 'traverse fibrils' extending into the cytoplasm and nucleoplasm from the PC. Magnification: × 100 000. Photograph kindly supplied by Dr G. G. Maul and reproduced by permission of the Editorial Board of *Int. Rev. Cytol.*

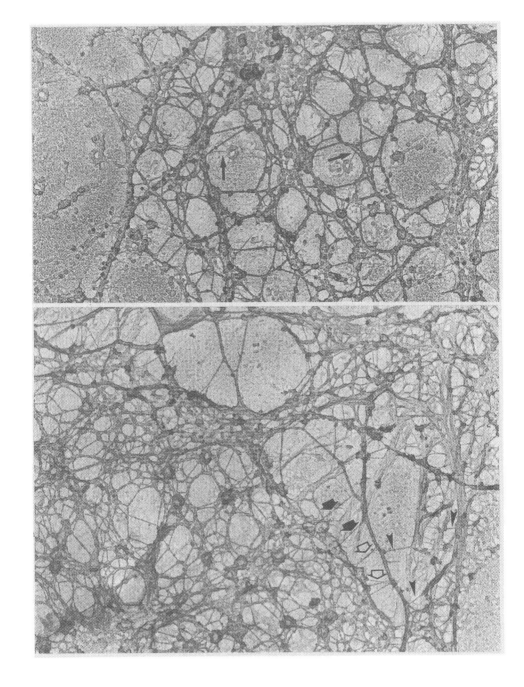

through the PC. A recent and well-supported case for this position was given by Carmo-Fonesca *et al.* (1987) (Figure 3.4). Many workers in the field have described a central structure in the PC lumen. This object has been variously perceived as a granule, a fibril and a tubule, present in some but not in all PCs, and there are divergent views about whether it is attached to the annular subunits (or the membrane margins) by fibrils or 'struts', or by some sort of diaphragm, coplanar with the membranes, or whether it is 'free-floating'. Others suggest that this object is simply particulate material which happened to be in transit between nucleus and cytoplasm when the specimen was fixed, or that it is a staining artefact; in short, that the PC itself does not have a central structure. This latter opinion seems to be held by a majority, though not a consensus, of workers in the field. Arguments against it include the general correspondence in size between the 'central granule/tubule' and the patent pore diameter determined by biophysical measurements (Paine *et al.*, 1975; Lang and Peters, 1984), and some recent biochemical advances (see next subsection).

Occasionally, the entire PC has been dismissed as an artefact of electron microscopy; for example, Archega and Bahr (1985) argued that the 'PC' results from association of smooth intranuclear toroidal structures called 'nuclear rings' with the NE. Although this view has not been accepted, the evidence adduced in its support might hint that PCs *in vivo* have a transient existence in the NE plane (Figure 3.5), and are simply 'snapshots' of structures assembled within the nucleus and in the process of movement along the fibrillar network through the NE to the cytoplasm (for aspects of this position, c.f. Kirschner *et al.*, 1977; Maul, 1982; and Carmo-Fonesca *et al.*, 1987).

Ultimately, these uncertainties will only be resolved by more complete biochemical characterization of the PC. Progress with this characterization is discussed on pp. 61–66.

3.1.2 Other aspects of ultrastructure

The ONM and INM themselves are structurally unremarkable. The ONM bears ribosomes, and generally resembles rough ER. General biochemical similarity between the ONM (if not the INM) and the ER is indicated by: (a) SDS-PAGE of the proteins of the two membrane systems, (b) cytochemical staining evidence, and (c) the fact that when the NE breaks down in late

---

Figure 3.4 Resinless section of the 'nuclear matrix-intermediate filament' system showing the fine fibrils interconnecting the PCs and the intermediate filament fibrils. Magnification: 7000. Photograph kindly supplied by Dr M. Carmo-Fonesca and reproduced by permission of the Editorial Board of *Eur. J. Cell Biol.*

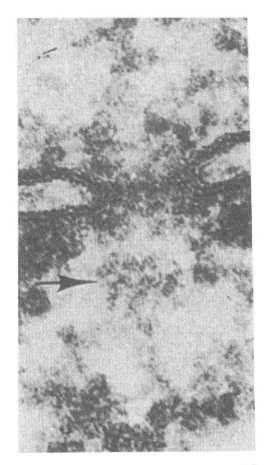

Figure 3.5 Fibrils extending into the nucleus seem to connect PCs with objects that might be nascent PCs within the nuclear compartment, suggesting that the PC might be a transient representation in the plane of the NE of a continuously-moving nucleocytoplasmic fibrillar system. Magnification: X 150 000. Photograph kindly supplied by Dr G. G. Maul and reproduced by permission of the Editorial Board of *Int. Rev. Cytol.*

prophase, the nuclear membranes and the ER form a single, morphologically undifferentiated pool of vesicles, from which the membranes of the daughter nuclei are reassembled during telophase. These very general similarities do not entitle us to describe the nuclear membranes and the ER as 'identical' (Richardson and Agutter, 1980). The apparent continuity between the ONM and the ER, and therefore between the perinuclear cisterna and the ER cisterna, provides a possible route for exchange of material between the nuclear periphery and the cell surface, or perhaps between nucleus and cytoplasm, but to date there is no substantial evidence that this route is actually used by the cell. However, some nuclear membrane components (e.g. the membrane glycoproteins – as opposed to PC glycoproteins, of which more later) are located on the cisternal faces of the nuclear membranes (Virtanen, 1977). Perhaps lateral flow through a Golgi/ER/nuclear membrane continuum puts these components in the right place and the right orientation.

In some cells under some conditions, e.g. melanocytes influenced by melanocyte-stimulating hormone, large outpocketings or 'blebs' form in the NE, or sometimes just the ONM (Turner *et al.*, 1979). These often protrude several microns into the cytoplasm. It could be that the sudden increase in transcriptional activity induced by the hormone effects changes in the nuclear surface that might be relevant to nucleocytoplasmic transport. However, although a pinching-off of these blebs would certainly add material of nuclear origin to the cytoplasmic compartment, albeit in membrane vesicles, there is no evidence that this actually happens. The 'blebs' might result from a combination of: (i) hormone-stimulated activation and expansion of parts of the peripheral heterochromatin, in which the genes are otherwise transcriptionally inactive, (ii) a resultant destabilization of the lamina, and perhaps (iii) membrane flow from ER to ONM. The point about 'destabilization of the lamina' is that the lamina is responsible for the structural integrity of the NE during interphase; the evidence for this, and its implications is discussed in section 3.3.

Because the lamina is so important in maintaining the nucleocytoplasmic boundary, it seems surprising that it was not clearly recognized until 17 years after the PCs were first described. The first clear electron-microscopic demonstration of the lamina *in situ* was by Fawcett (1966). The delay

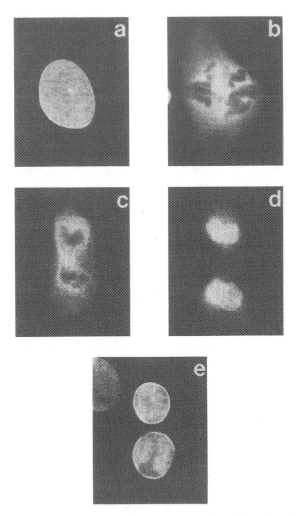

Figure 3.6 Fluorescence micrographs showing staining of interphase lamina and of dispersed lamins during mitosis. Magnification: × 480. Photographs kindly supplied by Dr L. Gerace and reproduced by permission of the Editorial Board of *J. Cell Biol.*

probably happened because the lamina is obscured *in situ* by the perinuclear heterochromatin, which is usually continuous around the nuclear periphery except at the PCs. It is obvious·as a fibrillar layer only on high-resolution micrographs (Figure 3.6) of a quality that was rarely attainable before the mid-1960s.

3.2 ASPECTS OF NUCLEAR ENVELOPE BIOCHEMISTRY

3.2.1 The architecture of the lamina

The composition of the isolated rat liver PCLF (Aaronson and Blobel, 1975) is dominated by three polypeptides with molecular weights in the 60–70 kDa range. Immunofluorescence, immunoperoxidase microscopy and differential solubilization studies show that these three proteins are confined to the lamina and are not present in the PCs. In 1980, the name **lamins** was coined for them (Gerace and Blobel, 1980); the rat liver lamins were called A, B and C in order of decreasing molecular weight. This nomenclature is now universally applied.

It soon became apparent that while all NEs had laminae, the details of lamina composition varied somewhat with cell type. The three-lamin profile found by SDS-PAGE of rat liver PCLF is typical of mammalian and avian somatic cells, but is not universal in nature; amphibian somatic cells have up to four lamins, many invertebrates have two, and gametes such as amphibian oocytes have one. Two-dimensional gels revealed more information about these proteins. In rat liver, lamin B is substantially more acidic than lamins A and C. More generally, irrespective of the total number of lamins in a somatic cell nuclear envelope, both the acidic (B-type) and the more basic (A-type) seem to be represented; in oocytes, the single lamin seems to be 'B-type' (Krohne *et al.*, 1978; Gerace and Blobel, 1980; Krohne and Benavente, 1986).

In rat liver, lamins A and C are actually different gene products (Laliberté *et al.*, 1984), despite the very high degree of sequence homology between them (McKeon *et al.*, 1986). They might be the results of a gene duplication fairly early in evolution; but the advantage of the duplication is not clear, because the protein-chemical differences between lamins A and C seem to be functionally insignificant. But the distinction between B-type and A-type lamins is quite another matter. In rat liver (and other tissues), lamin B has some sequence homology with A and C (around 30%), but the differences are functionally important. Lamin B seems to be much more tightly associated with the INM than A or C: for example, the INM-derived membrane vesicles during mitosis still contain some B, but almost no A or C (Gerace and Blobel, 1980), and Blobel and his colleagues have recently identified a specific lamin B receptor, a 58 kDa integral protein, in the INM (Worman *et al.*, 1988; see also Senior and Gerace, 1988). Lamin B, unlike A and C, binds tightly to this protein ($K_d = 2 \times 10^{-10}$ M) and the binding of lamin B to lamina-depleted nuclear membranes is blocked by antibodies against it.

Together with the finding that lamins A and C can form larger and more stable homo-oligomers than lamin B, but that lamin B can form hetero-oligomers with A and C (Aebi *et al.*, 1986), the discovery of the lamin B-INM association has provided an important step towards understanding lamina architecture.

Another important observation, made as a result of isolation and sequencing of the genes, is that lamins A and C have a high degree of sequence homology not only with each other but also with the cytokeratins (McKeon *et al.*, 1986; Franke, 1987). A few remarks about intermediate filaments and their components might help to indicate the significance of this.

Intermediate filaments are found in some higher invertebrates and in most vertebrate cells. They seem to be responsible for anchoring the nucleus, but their other functions are uncertain. They are very resistant to extraction under non-denaturing conditions. In any particular cell they consist of one of five classes of proteins: vimentin, desmin, glial filament protein, neuro-filament proteins or cytokeratins. By far the most heterogeneous of these classes is the cytokeratins. There are some 20 different cytokeratins in epithelial cells – this excludes the 8 known α-keratins characteristic of hair-producing cells. All intermediate filament proteins have similar 3-dimensional structures: they comprise an N-terminal head, an approximately 300-residue-long 'rod' section and a C-terminal tail. The high α-helical content of the proteins is contributed by the rods, which contain three double-stranded coiled-coil regions (1A, 1B and 2) (Figure 3.7) with characteristic heptad repeats in their structure. These three regions are separated by non-helical sequences. The head and the tail have no α-helix and their sequences vary greatly between different intermediate filament proteins. Homology between the intermediate filament proteins is mainly confined to the rod and is most striking at the beginning of the 1A region, and at the beginnings and the ends of 1B and 2.

This homology extends to lamins A and C. In these lamins, the 1B region is longer by 43 residues than the corresponding region in intermediate filament proteins, but otherwise there is marked similarity. The N-terminal head region (26 residues) is rich in serine and threonine, like most intermediate filament protein head regions. The lamin tails are also rich in hydroxy-

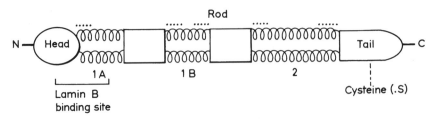

Figure 3.7 Schematic representation of an A-type lamin, showing the coiled-coil regions as parallel helices and the non-helical regions as oblong boxes.

aminoacids; and like some cytokeratin tails, they contain glycine-rich sequences (including the repeated tetrapeptides GGGS and VGGS), a predominance of acidic residues, oligohistidine sequences and cysteine residues (one in lamin C, five in lamin A; overall, lamin A differs from lamin C only in having a tail some 94 residues longer). Rather less is known about lamin B, but it is known not to contain the C-terminal oligohistidines or cysteines. There are two very striking cell-biological (as opposed to protein-chemical) differences between the lamins and the intermediate filament proteins: (a) the former are confined to the nucleus in interphase, the latter probably to the cytoplasm; (b) the former are ubiquitous in eukaryotes, and the latter are not.

How are the lamins linked together in the lamina? Presumably lamin A/C oligomers are attached to the monomeric INM-associated lamin B. The A/C lamins can polymerize under different ionic conditions, just as cytokeratins do, and they seem to form a square meshwork pattern *in vivo* with a 10 nm fibre width suggesting end-to-end linking of the molecules. If the B molecules join to A/C at the cross-over points of the mesh, the roughly 50 nm mesh spacing suggests that B is linked to A/C dimers *in situ*. A 32-residue N-terminal sequence of lamin A, synthesized from a cDNA, contains the binding site for lamin B. The binding of B to A and C seems to be co-operative, according to binding studies with column-immobilized lamins (Georgatos *et al.*, 1988). Presumably this co-operativity partly accounts for the stability of the interphase lamina (Franke, 1987). At mitotic prophase, the co-operativity is disturbed by phosphorylation of the binding domain of lamin B (Georgatos *et al.*, 1988).

Is the lamina simply an intranuclear continuation of the intermediate filament system, and is this the biochemical basis of the 'nuclear matrix/cytoskeleton' continuum identified by Penman and his colleagues (Capco and Penman, 1983; Fey *et al.*, 1986)? Indeed, lamin B and the intermediate filaments both seem to be attached presumably via the PCs, and there appears to be specific binding of intermediate filament proteins such as vimentin to lamin B (Georgatos and Blobel, 1987). However, the fibrils extending from the PC and apparently linking to cytoskeletal elements on one side and intranuclear filaments on the other are morphologically quite different (markedly finer) from intermediate filaments and laminae (Carmo-Fonesca *et al.*, 1987). They probably correspond to the traverse fibrils described by Maul (1982) and to the PC-radiating fibrils described earlier by Scheer and Franke (cf. Franke, 1974).

3.2.2 The biochemistry of the pore complex

One of the most disturbing aspects of studies on the PCLF for about a decade was the apparent absence of any uncontroversially identifiable PC component. Did this mean that the PC consisted of a large number of minor polypeptides, each present in quantities too small to visualize on conventionally-stained electropherograms; or that the major polypeptide components were lost during PCLF isolation? The former possibility seemed inconsistent with the highly regular, symmetrical ultrastructure of the PC; the latter seemed inconsistent with the fact that this ultrastructure is more or less maintained during isolation.

The breakthrough in the study of PC biochemistry came in 1982, when Gerace and his colleagues identified a monoclonal antibody against rat liver NE that reacted with a component located exclusively on the periphery of the PC (Gerace *et al.*, 1982) (Figure 3.8). This component proved to be a glycoprotein with an SDS molecular weight of 190 000. Being a large glycoprotein, this molecule (Gp190) stains rather poorly on gels with conventional protein stains, but can be identified on blots either with the monoclonal or with concanavalin A. Gp190 almost certainly corresponds to the large particle observed on the PC periphery by Unwin and Milligan (1982). It partially survives the procedures used to isolate the PCLF, which accounts for the maintenance of a recognizable PC morphology in the isolated material. Given its *in situ* location, Gp190 probably has a role in attaching the PC to the nuclear membranes. Recent sequencing studies support this view (Wozniak *et al.*, 1989).

Gerace and his colleagues went on to manufacture other monoclonal antibodies against PC components. Monoclonals that react specifically with the

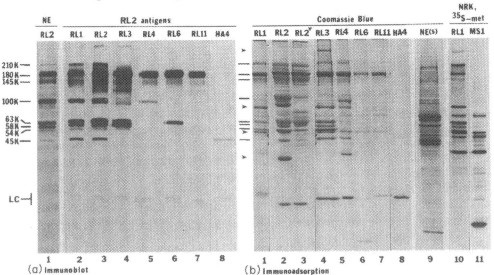

Figure 3.8 SDS-polyacrylamide gel electropherograms (a) Western blotted with the various monoclonals to reveal PC glycoproteins, (b) Coomassie stained. Photograph kindly supplied by Dr L. Gerace and reproduced by permission of the Editorial Board of *J. Cell Biol.*

PC as opposed to, say, the lamina give a characteristic immunofluorescence pattern: the surface of the nucleus shows a spotty fluorescence (Figure 3.9), not a continuous fluorescent band that is obtained when nuclei are treated with fluorescently-labelled anti-lamins (Davis and Blobel, 1986). To date, they have characterized a total of eight PC polypeptides other than Gp190 (molecular weights 210, 180, 145, 100, 63, 58, 54 and 45 kDa) using these monoclonals (Snow *et al.*, 1987). The 63 kDa component is almost certainly identical to the 62 kDa PC protein reported by Davis and Blobel (1986); the 145 kDa has an intranucleolar as well as a PC location; the 180 kDa, at least, is restricted to the nuclear face of the PC, showing that the biochemical organization of the PC is asymmetric along the nucleoplasm–cytoplasm axis. All eight components are glycoproteins, and they are all extracted to a greater or lesser extent by treatment with NaCl solutions at concentrations above 0.3 M. The fact that they are not visible on SDS-PAGE of PCLF can now be explained: (a) they stain poorly with protein stains; (b) they have largely been lost during the isolation procedure. Gerace and his colleagues calculated that each of these 8 proteins is present in quantities of 2–8 molecules per PC, in contrast to Gp190, of which there are about 25 molecules per PC.

The glycosylation of the PC proteins is an important issue. We normally think of glycosylation as being a Golgi-related process. Most extracellular and

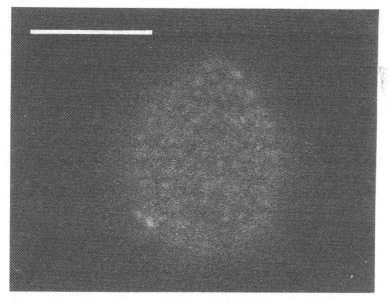

Figure 3.9 Fluorescence micrograph of antibodies against PC proteins, kindly supplied by Dr D. D. Newmeyer and reproduced by permission of the Editorial Board of *J. Cell Biol.* Contrast this punctate fluorescence pattern with the smooth continuous pattern obtained with anti-lamins (Figure 3.6). Bar = 10 μm.

cell-surface glycoproteins have their carbohydrate moieties added by Golgi enzymes: they contain N-acetylgalactosamine, and they react with such lectins as concanavilin A. The carbohydrate moieties of the PC glycoproteins are quite different. They contain N-acetylglucosamine (directly attached to a hydroxyl group on the protein) instead of N-acetylgalactosamine, and they do not react with concanavilin A; they all react with wheat germ agglutinin, which is specific for N-acetylglucosamine and sialic acid, and N-acetylglucosamine itself competes with this lectin, providing the basis for suitable control experiments. The enzymic basis of this unusual kind of glycosylation, which seems to be typical of cytosolic and especially of nuclear proteins, was described by Holt *et al.* (1987). The glycosylation occurs in the cytoplasm itself, not in the Golgi.

Most of the monoclonals developed in Gerace's laboratory cross-react somewhat (some react with all 8 components; one of them, RL11, is specific for the 180 kDa). However, there seems to be little or no sequence homology between any two of the PC proteins according to tryptic mapping evidence – though of course an epitope recognized by one of the monoclonals might extend only 1–2 residues on either side of the residue to which the carbo-hydrate is attached. Gerace and his colleagues conclude that the carbohydrate itself forms part of the epitope; an alternative is that the carbohydrate maintains the neighbouring polypeptide chain in a conformation that the immunoglobulin can recognize.

One of the PC proteins identified by Gerace's monoclonals, the 62–63 kDa, has been much studied because of its probable key role in nucleo-cytoplasmic transport (see Chapters 4 and 5 for further discussion). Its cytoplasmic precursor has been identified, and the molecular weight of this precursor is slightly less than 62 kDa. Precursor forms of membrane-associated proteins are often a little larger than the mature proteins because of the signal peptide, which is normally removed once the protein is incorporated in the membrane; but in the case of the 62 kDa PC component, the newly-translated precursor differs from the mature molecule in lacking its carbohydrate moiety. Glycosylation occurs within about 10 min of the end of translation, but incorporation into the PC is not necessarily immediate. According to Laura Davis (personal communication), the $t_{1/2}$ of incorporation is more or less equal to half the cell cycle time – just as we would expect from Maul's conclusion that PCs enter the NE only over the period of a whole cell cycle (cf. Maul, 1977).

All this information is potentially very valuable in the elucidation of nucleocytoplasmic translocation processes, but it does not add up to a detailed structural model of the PC: the glycoproteins probably make up only 5–10% of the mass of the PC. However, two other facts about them might be pertinent. First, the monoclonals all react with the **annulus**, according to immuno-electron microscopy performed with gold-labelled second antibodies (Snow *et al.*, 1987); but secondly, wheat germ agglutinin seems to react with the **central granule** or tubule (information from David Goldfarb). The apparent diameter of this putative structure seems to vary with the metabolic state of the cell: the more likely the PC is to be active in nucleo cytoplasmic transport, the greater the diameter of the central structure. The range of diameters observed on specimens flash-frozen in liquid nitrogen is from 9 to about 20 nm (note that the smallest value corresponds to the patent pore radius of 4.5 nm measured by biophysical techniques: see section 2.1.1). Perhaps the 'diaphragm' or 'spokes' that some electron microscopists have believed to occupy the PC lumen is/are the carbohydrate moieties of the glycoproteins, extending from the annulus where the polypeptide itself is located towards the centre of the structure; thus the lectin binds nearer to the centre of the PC than do the monoclonals.

These results lend weight to the growing body of opinion that PCs are contractile structures containing actin and myosin. There has been a fairly long history of observations suggesting that some form of myosin is present in the PC, but the possibility of myosin-based contractility seems to date back only to a paper by LeStourgeon (1978). Since then, some very convincing evidence for the presence of a form of myosin in the PC has accumulated, based on immunological and peptide-mapping studies; the case for actin is credible, but not yet so well established (Berrios and Fisher, 1986; Schindler

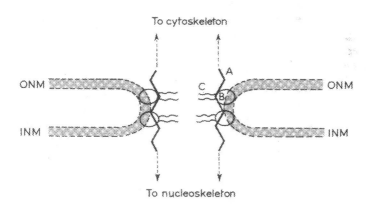

Figure 3.10 Schematic representation of a possible contractile structure in the PC. The polypeptide backbones of the PC glycoproteins (B) are membrane proteins associated with the ONM/INM junction, and they also bind to a contractile (actin-myosin?) framework extending, as 'traverse fibrils', into both cytoplasm and nucleoplasm (A). The carbohydrate moieties of the glycoproteins (C) extend into the PC lumen where they might appear as a diaphragm or struts attaching to a central granule. The contractile system is imagined as extending when energized, pulling the carbohydrates towards the membrane margin and opening the channel in the lumen.

and Jiang, 1986) (Figure 3.10). Obviously, contractility in the structures would be of the first importance in understanding transport processes, but at the moment we have no real idea how the proposed actin-myosin machinery and the PC glycoproteins are organized with respect to each other, what relationship either of them has to the traverse fibrils, or how any of these components actually functions in a translocation event. Perhaps it is worth noting that the PCs seem to be structurally stable only when they are attached to the lamina, so a detailed biochemical characterization of the attachment of lamin B to the PC might throw some light on the matter.

3.2.3 Some biochemical characteristics of the nuclear membranes

The molecular organization of the PC is obviously important for understanding nucleocytoplasmic transport. That of the lamina is also important, because of the role of the lamina in maintaining NE stability and because of its attachment to the PCs, which might be significant for PC function. But the relevance of nuclear membrane biochemistry is not so easy to see; in certain *in vitro* systems, the membranes can be removed with Triton without apparently perturbing at least some 'transport' processes (Agutter and Suckling, 1982; Schindler and Jiang, 1986).

Nevertheless, some Triton-soluble (therefore putative membrane) components do seem to have a function in translocation processes. One of them is a nucleoside triphosphatase (NTPase; an ATPase with a rather broad substrate specificity) with an SDS molecular weight of around 45 kDa. Cytochemical studies (Vorbrodt and Maul, 1980) indicate that this enzyme is located at the nucleoplasmic face of the INM; solubilization in Triton is more efficient when the isolated NE has been maintained in a sulphydryl reducing environment, e.g. by addition of dithiothreitol, suggesting that oxidative cross-linking to the lamins might otherwise occur. There is considerable evidence implicating the NTPase in mRNA translocation (Agutter, 1988) (Chapter 5). Another such component, also potentially implicated in mRNA translocation, is a 110 kDa protein that seems capable of binding messengers, at least those with poly(A) tails (Agutter, 1985b). Once again, this protein is accessible from the nucleoplasmic rather than the cytoplasmic face of the NE (Chapter 5), suggesting that it is located in the INM.

The INM is an unusually rigid membrane which seems to house most or all of the inositol-containing phospholipids of the NE (Agutter and Suckling, 1982 and discussion in Chapter 2). The inositol phospholipids and their metabolites seem to modulate the NTPase and perhaps other NE functions relevant to translocation processes (Smith and Wells, 1984). The details are discussed in section 5.23; the important point here is that so far as translocation processes are concerned, the PCs cannot be regarded as independent entities; INM located components, and possibly the lamins, seem important for their functioning as well as their structure.

3.3 BREAKDOWN AND REASSEMBLY OF THE NUCLEUS DURING MITOSIS

The nucleus and cytoplasm are entirely distinct during interphase, but apparently not distinct at all during mitosis in higher eukaryotes. Not surprisingly, most nuclear proteins redistribute throughout the cell at this time – though interestingly, mRNA does not (Rao and Prescott, 1967),

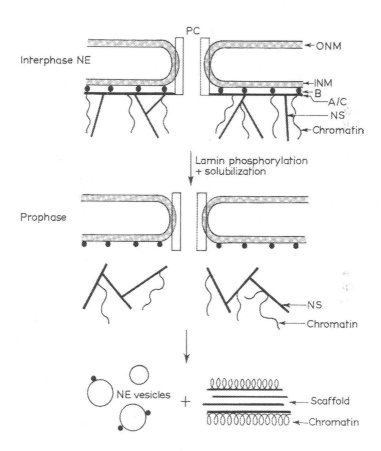

Figure 3.11 Diagrammatic summary of the studies of Newport and his colleagues. In the interphase NE, lamin B monomers are attached to A/C oligomers and to the membranes (for clarity, only contact of lamin B with the INM is shown). The lamina is also attached to the nucleoskeleton (NS) and the chromatin. When the lamins are phosphorylated at prophase, A and C are solubilized and the NS and chromatin subsequently detach. Much of lamin B remains membrane-associated. At the end of prophase, the NS and chromatin condense into chromosomes (NS material is believed to contribute to the chromosome scaffold) and the nuclear membranes vesicularize, with lamin B still attached to some of them.

another piece of evidence supporting the solid-state model of mRNA transport. Nuclear envelope breakdown and reformation during mitosis are very pertinent to the subject of this book.

Gerace and Blobel (1980) found that during mitosis the lamina breaks down and the lamins become randomly distributed through the cytoplasm, though some lamin B remains associated with the vesicles derived from the nuclear membranes. This breakdown seems to be caused directly by a hyper-phosphorylation of the lamins, catalysed by an endogenous (and so far unidentified) kinase which is stimulated by a component present in metaphase cytoplasm (Newport and Spann, 1987) (Figure 3.11). The phosphorylation sites seem to be located in the regions of the lamins that bind other lamins (page 60). This dissolution of the lamina is necessary for nuclear membrane vesicularization. (Hence the suggestion made earlier in this chapter that 'blebbing' of the NE results in part from local destabilization of the lamina.) But is lamin depolymerization a sufficient condition for vesicularization? And is it causally related to condensation of the chromosomes?

These questions have been answered by experiments with amphibian oocytes. When the oocyte is fertilized, it goes through a succession of divisions but little or no protein synthesis occurs until the mid-blastula transition, by which time the embryo comprises thousands of cells. Obviously the total volume of nuclear material in the blastula (not to mention the total nuclear surface area) is considerably greater than the volume (and surface area) of the oocyte nucleus. A net increase of nucleus without concomitant protein synthesis implies that the oocyte cytoplasm must contain pools of nuclear components waiting to assemble and organize in response to a suitable trigger. The studies by Kirschner and his colleagues (Forbes *et al.*, 1983) which was mentioned briefly in Chapter 2 revealed that a suitable trigger is naked DNA. Synthetic nuclei can therefore be made by adding DNA to oocyte cytoplasm – the intact cell is not necessary – and their dynamics of assembly and disassembly can be monitored when mitotic cytoplasm is added. These experiments have been conducted by Newport, and the publications describing them (Newport, 1987; Newport and Spann, 1987) are likely to become classics: they have opened the door to biochemical analysis of the mechanism of mitosis.

The results show that lamin depolymerization is not a sufficient condition for membrane vesicularization; the membranes only vesicularize after the lamina has broken down if a particular (presumably protein) factor is present in the mitotic cytoplasm. Also, chromosome condensation follows the dispersal of the lamina but is not directly caused by it. This process depends on the activation of DNA topoisomerase II. This ATP-dependent enzyme separates strands of chromatin that have become twisted through each other – obviously a precondition for the separation of the daughter chromatids at anaphase – and it is one of the two major components of the 'chromosome scaffold', the structural proteinaceous core of the metaphase chromosome (Marsden and Laemmli, 1979). (The other major component of the scaffold, known as ScII, is not characterized.) Therefore at least three mitotic cytoplasmic factors are necessary: the lamin kinase, the membrane vesicularization factor and the chromosome condensation factor; but lamina depolymerization is the initial event. The elusive 'maturation promoting factor' present during mitosis seems to activate these three factors rather than to participate directly in the process itself.

The order of events in assembly of the synthetic nuclei is largely a reverse of that in disassembly, but both processes are energy-dependent. The naked DNA first assembles into nucleosomes and DNA topoisomerase II-dependent chromatin condensation follows. Assembly of the lamina on the surface of this chromatin seems to involve some cytoplasm-induced modification, possibly the binding of a lamin receptor, and partial decondensation. The membranes are reassembled from vesicles in conjunction with the nascent lamina. ATP, and more especially analogues such as γ-S-AT, inhibit reassembly of the lamina, presumably because this process requires dephosphorylation of the lamins (Burke and Gerace, 1986). Most interestingly, antibodies against the lamins, microinjected into mitotic cells, arrest division at a stage when the daughter nuclei have a definite telophase configuration, but the chromosomes remain condensed and no nucleoli form (Benavente and Krohne, 1986). This implicates the lamina in chromosome condensation and decondensation processes, and therefore presumably in the dynamics of the NS and chromosome scaffold (page 22).

What does all this tell us about nucleocytoplasmic distributions of proteins (and perhaps other macromolecules)? Of course, the permeability barrier between nucleus and cytoplasm is temporarily lost during mitosis, so there is less restriction of movement of diffusible molecules, even large ones with karyophilic signals. Also, the changes in NS and chromatin organization are very dramatic, so the intranuclear binding sites of many proteins are likely to be radically altered (typically, lost). Therefore, there is a good deal of mixing between nucleoplasm and cytoplasm. The interphase distributions must be restored during or very soon after telophase, and this must involve the re-establishment of: (a) interphase intranuclear binding sites, and (b) the NE's characteristic permeability properties. DNA topoisomerase II and the lamins seem to play crucial parts in both these processes.

3.4 NUCLEOSKELETON, CYTOSKELETON AND NUCLEAR ENVELOPE

If the NS exists, then it obviously plays an important part in nuclear assembly and disassembly and in the re-establishment of intranuclear binding sites after telophase. It is likely that the NS and the metaphase chromosome scaffold share at least some core polypeptide components, one of which may well be DNA topoisomerase II itself (Berrios *et al.*, 1985); but we cannot be certain of this in a structure that is not well characterized.

The repeated use of the phrase 'well characterized' needs explanation. A method is described as well characterized when the following criteria are met: it gives reproducible results; the results correspond to those given by other relevant methods, are essentially free of artefacts, and relate in an obvious way to our understanding of the field; and the system used simulates the physiological situation in all relevant ways. A supramolecular structure is described as well characterized when it is possible to obtain isolated preparations that correspond reproducibly, in morphology and function, to the structure *in vivo*; when a fairly detailed knowledge of its molecular architecture is available – that is, not only its composition, but the spatial relationships and interactions between its major components; and when its dynamics are at least partly understood.

In the case of (say) microfilaments, there is a very substantial literature on isolation procedures, morphology, and behaviour. The core components have been identified; for the major proteins, we have sequence information, isolated and cloned genes, and monoclonal antibodies that recognize the proteins on Western blots and can be used in, for example immunofluor- escence and immuno-electron microscopy both *in vivo* and *in vitro*. (Nevertheless, there are hosts of other proteins, not all well characterized, that are more or less firmly associated with the fibrils.) The conditions and kinetics of assembly–disassembly processes are well known and, together with the biochemical evidence, throw light on the molecular architecture of the structure. Finally, there are at least reasonable hypotheses about assembly–disassembly mechanisms *in vivo*. Similar claims can be made about the other cytoskeletal elements, i.e. microtubules and intermediate filaments (though the functions of the latter are still quite unclear), about the lamina, and about nucleosomes.

The situation with regard to the NS is less satisfactory. There may be a morphologically acceptable isolated preparation (the nuclear reticulum); and there is some information – not obviously interpretable – about functions associated with the nuclear matrix. As for the major protein components, actin (Nakayasu and Ueda, 1984; Bladon *et al.*, 1988) and perhaps DNA topoisomerase II (Berrios *et al.*, 1985) are likely candidates, but there may be others. Labelled antibodies have been attached to either the *in situ* NS or an isolated nuclear reticulum (Chaly *et al.*, 1984; Berrios *et al.*, 1985), but antibodies have not yet shown any important organizational similarities between the *in situ* and *in vitro* structures. The ill-characterized status of the NS is certainly a stumbling block in the nucleocytoplasmic transport field.

Despite these gloomy reflections, one thing that does seem clear about the

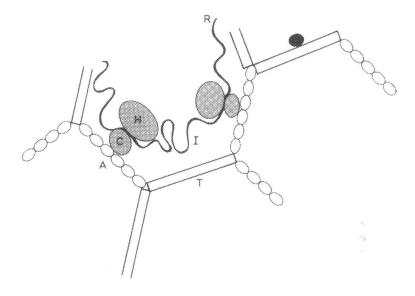

Figure 3.12 Schematic diagram showing possible mode of attachment of HnRNP particles to the NS. Two sorts of fibrils are envisaged, linked together in an anastomosing network: actin fibrils (A) and DNA topoisomerase II-containing fibrils (T). The HnRNP core particle (H + C) binds to the actin strands via the C-group proteins; the HnRNA (R) is attached to this particle. I = intron (introns are known to be located between rather than on the particles: Pederson, 1983).

nuclear reticulum (and by implication the NS: see Chapter 5) is that HnRNP is integrated into its structure. Some snRNPs seem to be similarly integrated (Habets *et al.*, 1985). This may have implications for the nature of RNA binding to the NS; needless to say, these implications cannot be extended to karyophilic proteins, if indeed any of them do bind to NS elements. Two methodologically distinct nuclear reticulum preparations which contain effectively all the HnRNP of the nucleus are disrupted by ribonucleases (Fig. 3.12) (Comerford *et al.*, 1986; Fey *et al.*, 1986); the anastomosing character of the fibrillar system is lost on this treatment. Interestingly, this effect is simulated by F-actin destabilizers (Comerford *et al.*), suggesting that the HnRNP is attached to the actin filaments of the reticulum. Similarly, there is a large body of evidence implicating actin (microfilaments) in the cytoskeletal binding of translationally active polysomes (see Jones and Kirkpatrick, 1988 for a clear summary of this evidence). According to the studies of van Eekelen and van Venrooij (1981) on isolated nuclear matrices, HnRNP is attached to the matrix and therefore perhaps to the reticulum and the NS via its core C-group proteins (the six core polypeptides of HnRNP are regarded as three pairs, A, B and C, with molecular weights ranging from 31 to 43 kDa).

In the nucleolus, there seems to be a residual fibrillar component that provides the framework for transcription and processing of the ribosomal genes and assembly and nucleocytoplasmic transport of the ribosomal subunits. Three proteins have been fairly definitively localized in this structure: a 180 kDa protein involved in transcription and processing (Schmidt-Zachmann *et al.*, 1987), the 38 kDa protein B23 which probably functions in particle assembly (Chan *et al.*, 1986; Fields *et al.*, 1986), and a 145 kDa protein that might function in storage and transport (Krohne *et al.*, 1972). In the protozoon *Tetrahymena pyriformis* there is some evidence that Ca^{2+} and Mg^{2+} modulated contractility of the structural framework of the macronucleus is directly relevant to ribosome subunit transport (Wunderlich *et al.*, 1984) and possibly to the transport of other RNAs. This might suggest that the earlier discussion about PC contractility should be extended to the NS; a possible common component is actin.

Actin can perhaps be regarded as the major characteristically eukaryotic structural protein. It seems to be present in every eukaryotic cell and it has yet to be demonstrated in a prokaryote (except one into which an actin gene has been engineered). Given that the separation of nucleus and cytoplasm is the defining characteristic of a eukaryotic cell, which in a sense makes nucleocytoplasmic transport the primary distinctive eukaryotic process, it is tempting to combine these general observations in the speculative claim that actin is fundamental to all nucleocytoplasmic macromolecule transport machineries. This theme will be resumed in Chapter 6.

3.5 OVERVIEW

The structures that are most directly relevant to the study of nucleocytoplasmic transport, the PCs, are octagonally symmetrical cylinders that pass through the planes of both nuclear membranes and have fibrous connections with the cytoskeleton and presumably the NS. The annulus (the cylinder wall) contains: (i) a form of myosin, (ii) (probably) actin, and (iii) at least eight glycoproteins with apparent molecular weights ranging from 45 to 210 kDa. There is some controversy about the significance of the 'central granule/tubule' visible in the centre of the PC lumen in many electron micrographs, but recent evidence suggests that the N-acetylglucosamine-containing carbohydrate moieties of the glycoproteins are inwardly-directed and might constitute this 'structure'. The apparent contractility of the 'central granule/tubule', which seems to be pertinent to translocation of material across the NE, might be a manifestation of the relative movements of these glycoprotein regions resulting from actin–myosin movements. The outer part of the PC annulus contains about 25 molecules of a different kind of glycoprotein (Gp190), which reacts with concanavilin A, not with wheat germ agglutinin as the 'inner' PC glycoproteins do. Gp190 is probably important in attaching the membranes to the PC.

The lamina consists of 1–4 proteins known as lamins (the number depends on the type of cell and the species) which are encoded on genes belonging to the intermediate filament multigene family. Unlike intermediate filament proteins, the lamins are exclusively nuclear during interphase; and unlike intermediate filament proteins they are ubiquitous in eukaryotes. Despite these differences, they show very marked sequence homology and structural similarity with the cytokeratins. In mammalian and avian somatic tissues, which contain three distinct lamins, the most acidic (lamin B) is apparently most directly associated with the INM, and probably serves as the attachment site for oligomers of the other, more markedly cytokeratin-like lamins (A and C). Breakdown of the lamina seems to be a necessary though not sufficient condition for nuclear membrane vesicularization and for chromosome condensation at mitotic prophase; and at telophase, reformation of the lamina is a precondition for chromosome decondensation and re-establishment of the nuclear membranes. Lamina breakdown and reformation is dependent on endogenous phosphorylation and dephosphorylation of the lamin–lamin binding sites.

The relevance of the lamina to chromosome condensation and decondensation probably reflects the dependence of NS integrity on it, the NS and its presumed mitotic counterpart (the chromosome scaffold) being important for chromatin organization as well as ribonucleoprotein attachment. The NS is not yet adequately characterized, but its major proteins apparently include actin and DNA topoisomerase II.

The main issues that await resolution in this area of research are: the biochemical nature of the associations between the lamina and (a) the NS, (b) the PCs and (c) the INM; the detailed architecture and the dynamics of the PC; and the nature and significance of the 'traverse fibrils' that link the PCs to the cytoskeleton and probably the NS. It should also be emphasized that the precise nature of protein and RNA binding to the NS (and the cytoskeleton) is not known.

4 Nucleocytoplasmic protein distributions

4.1 KARYOPHILIC PROTEINS

Nucleocytoplasmic protein distributions are randomized during mitosis. During telophase, most proteins and other soluble macromolecules are kept out of the nascent nuclear compartment, apparently by gel exclusion mechanisms (Section 3.3; Beck, 1962; Swanson and McNeil, 1987). It follows that the high nucleus : cytoplasm concentration ratios of most or all karophilic proteins are established during interphase.

Bonner (1975a,b) collected and analysed a mass of data from a series of microinjection experiments on frog oocytes. He drew the following conclusions:

1. Not surprisingly, the oocyte nuclei took up small proteins and excluded large ones. The effect of size was compatible with the results of Feldherr's earlier colloidal gold studies. It also matched the patent pore radius value which Paine and his colleagues published more or less contemporaneously.
2. However, some proteins accumulated in the nuclei, and this effect did not seem to be related to molecular size. Histones, for example, accumulated very rapidly in the nucleus after cytoplasmic microinjection.
3. When labelled oocyte cytoplasmic proteins were microinjected, they mostly remained in the cytoplasm of the recipient cell; and when labelled oocyte nuclear proteins were microinjected, they mostly entered the recipient cell's nucleus.

Bonner's third conclusion suggested that nuclear proteins have 'homing devices': intramolecular signals that make them find and enter the nucleus, presumably by means of specific transport processes associated with the PCs.

His second conclusion suggested that accumulation in the nucleus (**intra-nuclear binding**) has to be distinguished from uptake into the nucleus.

4.1.1 Intranuclear binding: the principles

In principle, we could explain high nucleus : cytoplasm concentration ratios of proteins by either of two models: (a) active transport through the NE, maintaining a higher activity of the (soluble) protein in the nucleoplasm than in the cytoplasm; (b) intranuclear binding, maintaining a higher total concentration (soluble + immobilized) in the nucleus but more or less equal activities in the two compartments. Feldherr's NE disruption studies (Section 2.1.3), the observations of Beck (1962), and other lines of evidence to be discussed in this chapter, all indicate that model (b) applies in practice. The more tightly a protein binds to solid-state elements in the nucleus (or cytoplasm), the fewer molecules there are in aqueous solution in that compartment, so the lower its activity. The more completely a protein is excluded from some fraction of the cytoplasmic (or nuclear) volume, the higher its activity in the remaining volume. Thus, very high nucleus : cytoplasm concentration ratios can be explained in terms of cytoplasmic exclusion and intranuclear binding.

This argument applies to any transportable material, not just to proteins. For instance, Horowitz and Paine (1976) showed that the very different Na^+ and K^+ concentrations of oocyte cytoplasm and nuclei were entirely attributable to binding and exclusion effects: there are no activity differences between the compartments for either ion, and there is no ionic potential across the NE. But if all asymmetric protein distributions are to be explained by this general principle, can we say anything more specific, or more useful, about the actual mechanisms involved?

Our present knowledge suggests the following classification of nucleus-located proteins (a tendency to propose classification schemes is a characteristic trait of biologists):

1. DNA binding proteins:
   (a) histones and other general DNA-binding proteins, e.g. the HMGs;
   (b) sequence-specific binding proteins such as large-T (see below).
2. RNA binding proteins:
   (a) the core polypeptides of HnRNP;
   (b) the protein components of the SnRNPs;
   (c) ribosomal proteins and proteins specifically bound to ribosomal precursors.
3. Structural proteins:
   (a) the lamins;

　(b) the putative NS core proteins, e.g. actin and DNA topoisomerase II;

　(c) nucleolar structural proteins.

4. Others:

　(a) enzymes such as nucleic acid polymerases, and other proteins e.g. nucleoplasmin, that bind to proteins in groups 1–3;

　(b) miscellaneous.

The principle underlying this rather crude scheme is that proteins can accumulate in nuclei by virtue of binding to DNA, to RNA or to other proteins that are themselves nuclear. For some classes, such as 1(a) and 2(b), the relevant binding sites have been characterized and the nature of the intranuclear binding is correspondingly well understood. For others, e.g. 3(b), characterization is far from complete. Class 4(b) is simply an indication that the list of nuclear proteins that we can name, and describe at least in outline, is the tip of a rather dense iceberg.

Modification of a protein within either cytoplasm or nucleus could alter its affinity for a binding site, thus altering its activity within the compartment and greatly changing its nucleus : cytoplasm concentration ratio. Protein phosphorylation is an obvious example: it could change nucleocytoplasmic distributions during different stages of the cell cycle or in response to hormonal or other signals (Chapter 6). The discovery of an intranuclear lectin, which might be NS-associated given its affinity for DNA polymerase and HnRNPs (information from Maurice Monsigny), suggests that glycosylation could be another way of altering the intranuclear binding properties of proteins; but this is mere speculation. Equally, the dynamics of the cytoskeleton, and presumably the NS, could make both the availability of binding sites and the efficacy of gel exclusion change with the stage in the cycle or the metabolic state of the cell; this would similarly alter the protein's nucleocytoplasmic distribution.

4.1.2 Intranuclear binding: some evidence

There are data for the three proteins believed to be involved in nucleosome assembly – N1, N2 and nucleoplasmin – that convincingly demonstrate the importance of intranuclear binding for their high nucleus : cytoplasm ratios during interphase.

1. Feldherr's NE puncturing studies, described in Chapter 2, have been mentioned several times already.
2. Nucleoplasmin's intranuclear diffusion coefficient is much lower than that for an equivalent-sized dextran (Schulz and Peters, 1987).
3. When a *Xenopus* nucleus is isolated by microdissection under paraffin oil to prevent protein loss by diffusion, and is then fused to a similar-sized reference phase (in this case an agarose gel), none of the three proteins

shows a marked tendency to enter the reference phase even when the NE is punctured. Bovine serum albumin, which diffuses through the PCs extremely slowly, passes freely between nucleus and reference phase under these conditions when the NE is disrupted (Paine, 1987).

4. When the NE is completely removed from an amphibian oocyte, nucleoplasmin still accumulates in the nucleus when injected into the cytoplasm (De Robertis, 1983).

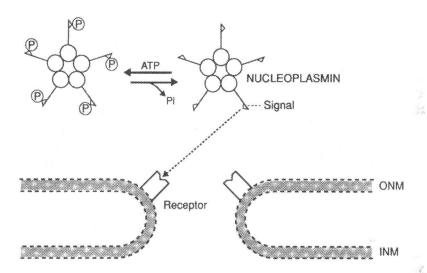

Figure 4.1   Schematic diagram showing the effect of phosphorylation on the binding of nucleoplasmin within the nucleus and its capacity for redistribution during mitosis.

Moreover, there is some evidence to show that the nuclear binding affinity can indeed be changed by protein phosphorylation (Figure 4.1). The release of nucleoplasmin from the *Xenopus* oocyte nucleus during NE breakdown at the onset of meiosis seems to be associated with phosphorylation of this protein; the analogy with lamina breakdown by phosphorylation is very striking (Sealy *et al.*, 1986).

4.2 NUCLEAR LOCATION SIGNALS

There is no doubt that intranuclear binding explains most of the asymmetric protein distributions between nucleus and cytoplasm during interphase. But we know from the passive permeability properties of the NE (Chapter 2) that average-sized proteins (40–50 kDa) diffuse only very slowly through the PCs by passive means, and many karyophilic proteins are of much more than average size. Yet nucleus:cytoplasm protein distributions are established rapidly *in vivo*. It is therefore necessary to postulate that such proteins have **nuclear location signals**, the 'homing devices' implicit in Bonner's data: parts of the polypeptide sequence that ensure rapid translocation through the NE, but need not have anything directly to do with intranuclear binding. Three years after Bonner's publications, in experiments deploying the then-novel technique of two-dimensional gel electrophoresis of proteins, De Robertis *et al.* (1978) obtained direct evidence for such signal sequences. Movement of some microinjected proteins into the nucleus was far too rapid to explain in terms of passive diffusion through the PCs, even allowing for the fact that the passive diffusion rate is a function not only of patent pore radius but also of the activity difference between the compartments.

There was nothing novel or surprising in the idea of 'signal sequences'; for example, the existence of **signal peptides** on proteins scheduled for migration through or into a membrane was fairly well established at the time and nowadays is undergraduate textbook knowledge. Precursor proteins (the immediate products of translation) are usually larger by 1–3 kDa than mature membrane or extracellular proteins. The size difference is attributable to oligopeptide sequences, typically but not universally at the N-terminus, removed by a 'signal peptidase' after the precursor has entered the correct compartment of the cell. Did the findings of De Robertis *et al.* (1978) mean that karyophilic proteins have some special kind of signal sequence, distinct from signals for passage through (say) the ER, but recognized by a receptor in the PC and removed once the protein is inside the nucleus?

This turned out not to be the case. Dabauville and Franke (1982) showed that proteins do not enter the nucleus as precursors: the immediate products of translation are not precursors, but the mature proteins. This conclusion is actually obvious from the fact, established by Bonner and others, that mature nuclear proteins from one cell migrate into the nucleus of another after injection into the cytoplasm. Therefore, if there are such things as nuclear location signals (a hypothesis further supported by the Dabauville–Franke experiments), they must be parts of the mature protein itself. Could there be some oligopeptide sequence that is common, perhaps in a few variant forms, to all nuclear proteins?

4.2.1 Nucleoplasmin

Nucleoplasmin, the most abundant karyophilic protein in oocytes and many other types of cell, is a large acidic pentameric molecule. Each subunit (the five are identical) has a molecular weight around 22 kDa and consists of two major domains separable by limited proteolysis: an N-terminal 'head' and a

C-terminal 'tail'. The head groups of the intact pentamer are directed towards the centre of the molecule. The intact pentamer seems to enter the nucleus *in vivo*. Because nucleoplasmin is so abundant and well characterized, and because it presumably must contain at least one of the putative nuclear location signals (nucleoplasmin-coated gold particles enter the nucleus; Chapter 2), it was the obvious protein in which to try to identify the relevant sequence.

More or less contemporaneously with the work of Dabauville and Franke, Dingwall *et al.* (1982) showed that the tail regions of nucleoplasmin contain the location signal(s). By varying the extent of proteolysis *in vitro*, forms of

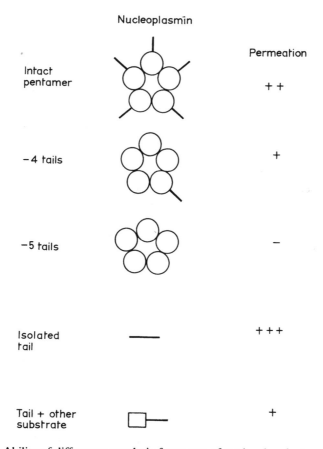

Figure 4.2 Ability of different proteolytic fragments of nucleoplasmin to permeate nuclei. Any other substrate (e.g. a colloidal gold particle) can permeate the NE if it is attached to the tail of nucleoplasmin which contains at least one nuclear location signal.

nucleoplasmin containing 4, 3, 2, 1 or no tails were prepared and were microinjected into amphibian oocytes (Fig. 4.2). Only the completely tailless derivative was excluded from the nucleus. Isolated tails entered the nucleus rapidly on microinjection; isolated heads could not enter at all. However, when the tailless 'core' of nucleoplasmin was microinjected directly into the nucleus, it remained there; there was no mechanism for excluding tail-less heads from the nucleus. Indeed, it appears that while the nucleoplasmin tail contains the nuclear location signal, the high nucleus : cytoplasm concentration ratio of this protein during interphase is to be explained by intranuclear binding of the heads, which contain polyglutamic acid tracts (probable histone binding sites: Dingwall *et al.*, 1988). The distinction between the two main aspects of karyophilic protein dynamics, uptake and accumulation, is clearly illustrated by nucleoplasmin.

4.2.2 SV$_{40}$ large-T antigen

However, the first nuclear location signal to be fully characterized was not the one in nucleoplasmin, which proved surprisingly elusive (see later), but in a completely different protein: the large-T antigen of Simian virus-40 (SV$_{40}$). This protein, which has a molecular weight in excess of 90 kDa, binds to the host's nuclear DNA and plays a key part in incorporating the viral into the host genome. It is intrinsically karyophilic, capable of entering the host cell nucleus alone, apparently unaided by DNA or any other viral or host cell component. When purified large-T is injected into a cell, it rapidly accumulates in the nucleus; almost none remains detectable in the cytoplasm (see Chapter 2 for the relevant methods).

During the early 1980s, A. E. Smith and his colleagues were studying the DNA binding site of large-T, partly to clarify some aspects of the host–viral genome interaction *per se*, and partly in the interests of characterizing another protein that bound to a specific DNA sequence. Specific DNA binding proteins were being identified and characterized for the first time during this period. Essentially, their experiments depended on manipulation of the isolated, cloned large-T genes; by removing or replacing portions of the gene, they could produce variant proteins with altered DNA affinities and specificities (and perhaps alterations in other functions). Among their experiments were some host-cell microinjection studies, and through these they found one variant protein that on microinjection would not enter the nucleus; it was restricted to the cytoplasm.

Accidents are sometimes said to play a big part in the history of science. It is more accurate to say that some advances in science have developed from timely accidents, the significance of which has been recognized by researchers with sufficient insight. Smith and his colleagues soon refined this chance observation into a provisional characterization of the first nuclear location signal: a highly basic oligopeptide sequence located in the N-terminal half of large-T, in the linker region between two highly structured domains of the protein (Kalderon *et al.*, 1984). Further studies made it apparent that replacement of one of the five key residues, lysine-128, with any other amino acid except arginine, was sufficient to give a completely cytoplasm-restricted variant; so a basic residue at position 128 was essential for nuclear permeation. Replacement of any of the other four residues gave variants that appeared in both nucleus and cytoplasm after microinjection.

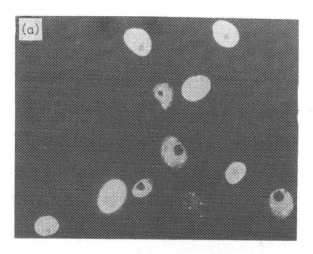

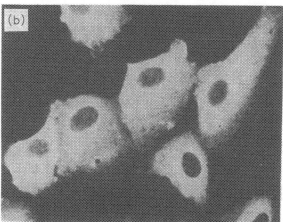

Figure 4.3   Fluorescence micrographs showing the ability of wild-type SV$_{40}$ large-T antigen to enter the host cell nucleus but the cytoplasmic retention of a mutant form of the protein in which lysine-128 is replaced by threonine. Magnification: × 500. Photographs kindly supplied by Dr A. E. Smith and reproduced by permission of the Editorial Board of *Proc. R. Soc. Lon. (Biol.)*.

Replacement of residues more or less distant from this sequence had no effect on nuclear location at all. However, DNA binding of large-T depends on a quite distant and distinct part of the protein. Once again, we have an example of the mechanistic distinction between uptake and accumulation: the nuclear location signal and the intranuclear binding domain can be unequivocally distinguished within a single molecule.

If this short peptide sequence is both necessary and sufficient for nuclear location, then another prediction can be made. If a fusion protein is made, in

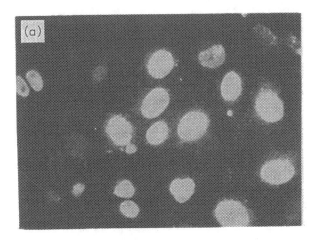

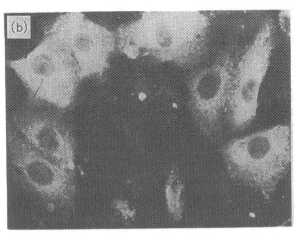

Figure 4.4   Fluorescence micrographs showing the uptake into nuclei of pyruvate kinase modified by addition of the wild-type nuclear location signal peptide from the large-T antigen, but not the mutant signal. Magnification: × 500. Photographs kindly supplied by Dr A. E. Smith and reproduced by permission of the Editorial Board of *Proc. R. Soc. Lond. (Biol.).*

which a normally exclusively cytoplasmic protein is supplemented with the putative signal, then the fusion protein should enter the nucleus upon microinjection. This prediction was tested with pyruvate kinase. Pyruvate kinase with the added oligopeptide duly entered the nucleus; pyruvate kinase manipulated in other ways, e.g. by addition of other short oligopeptides, retained its usual cytoplasmic location (Figure 4.3). In particular, addition of a defective SV_{40} signal sequence (lysine-128 replaced by threonine) generated a cytoplasm-restricted fusion protein. The case was clearly established (Figure 4.4).

Subsequent studies have shown that the more copies of the nuclear location signal there are in a protein molecule, the more rapid and extensive the accumulation in the nuclei; and different proteins with the same signal compete for entry into the nucleus (Goldfarb *et al.*, 1986; Lanford *et al.*, 1986; Roberts *et al.*, 1987; Lanford *et al.*, 1988). It is generally accepted that these studies have revealed the basis for protein translocation through the NE. However, the results discussed so far could (just) be consistent with a model in which the 'nuclear location signal' actually specified binding to a limited number of intranuclear sites, without determining the binding affinity, rather than interaction with a specific receptor in the PC. Finally, to confirm the relevance of these data to protein translocation through the NE requires another stage of argument. This will be considered further in Section 4.3.

4.2.3 Is the large-T signal universal?

Since the pioneering study of Smith's group, many nuclear location signals have been identified in karyophilic proteins. Many of them are strikingly similar to the one in SV_{40} large-T: highly basic oligopeptide sequences. However, some of them seem to form another class: short sequences in which (typically) two basic residues are separated by (typically) 2–3 neutral ones, the prototype of which is the yeast MAT-α_2 sequence first characterized by Hall *et al.* (1984). Some examples are shown in Table 4.1.

Table 4.1 Examples of large-T signal peptides

Protein	Location signal	Reference
SV_{40}	P_{126}–K–K–K–R–K–V–E	Kalderon *et al.* (1984)
MAT-α_2	R–P–A–A–T–K–K	Hall *et al.* (1984)
Polyoma large-T	(a) P_{280}–K–K–A–R–E–D	Peters (1986)
	(b) V_{533}–S–R–K–R–P–R	
N1	(a) V_{531}–S–R–K–R–P–R	Kleinschmidt and Seiter
	(b) A_{548}–K–K–S–K–Q–E	(1988)
Nucleoplasmin	R_{152}–P–A–A–T–K–K–A–G–	Bürglin and De Robertis
	N–A–K–K–K–K–L–D–K–E–D–E	(1987)

A possible unifying characteristic of these disparate sequences is a turn-helix-turn motif, with at least one basic face in the helix. It will be interesting to see if further evidence supports this generalization. If so, the large-T signal is universal not in primary sequence terms, but in secondary structural features. If, further, it can be confirmed that the signal sequence specifies binding to a NE receptor, then the proposed secondary structure motif would have interesting implications for the nature of signal-receptor binding.

Particularly interesting cases (not shown in the table) are the lamins, where immediately after the end of helical domain 2 in the rod, at the start of the C-terminal tail, the sequences

K	E	**K**	**R**	**K**	R	I
A	S	**K**	**R**	**R**	R	L
V	T	**K**	**K**	**R**	K	L

have been found in three different species (information from Frank McKeon). The similarity with the large-T signal is striking. The important point is that this sequence is absent from cytokeratins and other intermediate filament proteins. The triplets shown in bold type are apparently crucial for nuclear location: replacement with isoleucine or aspartate, especially at the third residue of this triplet, generates lamin variants that cluster around the nuclear surface but on the cytoplasmic side: in other words, they have the intracellular distributions of cytokeratins. These variant proteins form tubular structures which do not disassemble during mitosis, possibly because they are separated from the lamin kinase *in situ*. This tiny part of the lamin primary structure seems to be sufficient for the main biological difference, i.e. intracellular location, between lamins A and C and the cytokeratins.

4.2.4 Signals with different efficiencies

In a series of experiments by Lanford *et al.* (1986, 1988), the wild-type or mutant forms of the SV_{40} large-T nuclear location signal have been cross-linked to bovine serum albumin, ovalbumin, immunoglobulin G, ferritin and various other cytoplasmic or extracellular proteins. The cross-linker used was maleimidobenzoylhydroxysuccinimide (MBS) (Fig. 4.5). When the wild-type signal was used, all the proteins entered the nuclei and accumulated there after cytoplasmic microinjection. When a defective signal (one in which lysine-128 is replaced, for example by asparagine) was incorporated instead, nuclear uptake was abolished. Different substitutions at the lysine-128 position gave location signals with a range of efficiencies. In a protein with a wild-type signal, nuclear uptake took 15 min. When the lysine-128 was replaced by arginine, D-lysine and ornithine, nuclear uptake took 1, 3 and >6 hours respectively. Other substituents for lysine-128 resulted in nuclear uptake that was so slow as to be essentially unmeasurable.

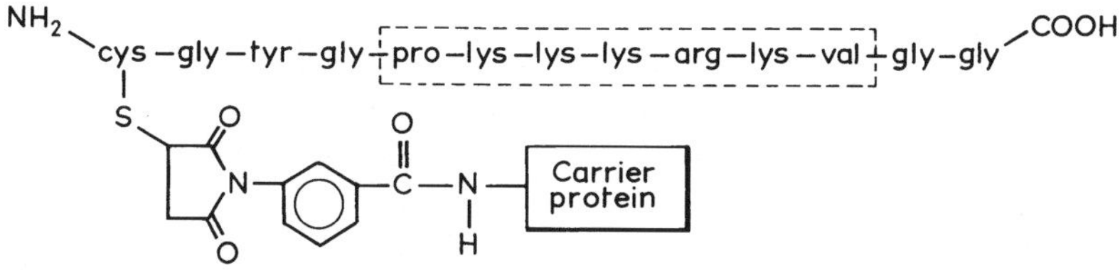

Figure 4.5 The MBS cross-linker used in the experiments by Lanford and his colleagues (see text). The SV_{40} large-T nuclear location signal is part of the peptide attached to the carrier protein, and is outlined in dotted lines.

When increased numbers of copies of the location signal peptide were linked to the same protein molecule, the nuclear uptake rate was increased correspondingly. Thus several copies of an inefficient signal can be as effective as just one or two copies of a really efficient one. Of course, the molecular sieving properties of the NE remain relevant: small proteins are taken up more rapidly by nuclei than big ones with the same location signal complements. More recent joint studies by Lanford's and Feldherr's laboratories (Dworetzky *et al.*, 1988) have confirmed that uptake rates of colloidal gold particles are increased by increasing the numbers of location signals and decreasing the particle size.

Finally, when Lanford and his colleagues incorporated the wild-type signal back-to-front (DEVKRKKKPC instead of vice-versa), there was no nuclear uptake. This last result demonstrates the specificity of the signal: the turn-helix-turn motif remains, but the distribution of positive charges about the helix has been altered. Studies by Goldfarb *et al.* (1986) using synthetic peptides incorporating location signals corroborated these findings and gave data from which a value for the maximum nuclear uptake rate could be calculated (Chapter 6).

4.2.5 Some caveats: or, how to control a bandwagon

The ease with which genes can be isolated, cloned, sequenced and manipulated nowadays has led to the identification of nuclear location sequences becoming something of a growth industry. A great many publications in the field are now devoted to identifying yet possible another permeation signal in yet another karyophilic protein. It is not difficult to generate papers by pursuing this line of investigation, irrespective of whether any interesting biological question is being answered.

Smith and his colleagues (Roberts *et al.*, 1987) have drawn attention to the fact that to identify a sequence similar to the large-T or the MAT-α_2 signal in a nuclear protein is not enough to show that the sequence in question actually functions as the nuclear location signal *in vivo*. Some workers seem content with this information, but the following points have to be considered as well.

1. Ablation or appropriate modification (e.g. replacement of lysine-128 in SV$_{40}$ large-T) of the putative signal must restrict the protein wholly or partially to the cytoplasm after microinjection.
2. The efficiency of a potential location signal depends on its protein context. The relevant residues have to be exposed at the protein surface, though this will usually be the case for highly basic sequences; in SV$_{40}$ large-T the signal seems to remain effective irrespective of where it is placed in the protein. However, in other cases the signal might function only if it is situated in its wild-type position, e.g. at one end of the molecule; if it is moved to another part of the polypeptide it might not work.
3. There can be more than one location signal in a protein, and they don't necessarily all have the same efficiency. Good signals dominate over poor ones (that is, poor signals don't inhibit uptake if a good one is present). As Lanford *et al.* (1986, 1988) showed, the effects of multiple signals are additive. For example, in the large-T antigen of polyoma virus (as opposed to SV$_{40}$), there are four potential location signals, at least two of which have to be present for efficient uptake into the nucleus. Both the likeliest-looking signals appear defective when compared with the SV$_{40}$ large-T signal (Table 4.1), so neither is sufficient alone.

   Having more than one location signal seems to be a common property of nucleus-targeted proteins. The main reason that the nucleoplasmin signal sequence remained so elusive for so long is that it actually comprises three adjacent or overlapping potential signal sequences: two MAT-α_2-like ones (residues 152–157 and 158–163) and a large-T-like one (residues 162–168: Table 4.1). Their combined efficacy is great. Bürglin and de Robertis (1987) published this composite sequence shortly before Dingwall *et al.* (1988), who had obtained the same information at more or less the same time.
4. A nuclear permeation signal might be modified *in vivo*, e.g. by phosphorylation, as Dreyer and her colleagues have suggested. This can render it inactive until the modification is reversed.
5. The effect of a nuclear location signal can also be overridden by sequences that specify binding to other cell components. For example, a membrane-binding domain will make the protein enter a membrane (e.g. ER) irrespective of the presence of a nuclear location signal. Removal or masking of such domains could be another way of regulating nuclear uptake of proteins.
6. Results from fusion proteins can only be trusted if the relevant control is performed: incorporation of a defective signal should result in the usual cytoplasmic instead of nuclear location upon microinjection.

There are some more caveats, not discussed by Smith and his colleagues. For instance, some proteins that can become nucleus-located *in situ* don't have a nuclear location signal at all. Tight binding to a protein that does have a location signal seems to be sufficient, as in the case of a monoclonal antibody against two human nuclear proteins known as IEF 8Z30 and IEF 8Z31 (Madsen *et al.*, 1983); obviously, the monoclonal does not have a location signal of its own. Similarly, when lamins A and C are crippled by replacement of the crucial basic residues in their location signals and the variant proteins are microinjected into an oocyte, some traces of them still enter the nucleus, though the comments made previously about these lamin variants remain essentially true. Frank McKeon suggests that these variants can enter by forming dimers or other small oligomers with endogenous normal (wild-type) lamins. If McKeon's explanation is correct, then the fact that cytokeratins never enter the nucleus indicates that they are incapable of forming hetero-oligomers with lamins A and C, despite their great structural similarities.

Yet another point is that the supposed 'all-or-nothing' character of nuclear protein uptake might be illusory. Many publications, including the pioneering one from Smith's group (Kalderon *et al.*, 1984) certainly imply that a normal location signal gives total nuclear uptake and a defective one gives total restriction to the cytoplasm. But the presence of different numbers of signals with different efficacies in many proteins, and Lanford's study of the effect of signal efficacy on the time-course of uptake as opposed to its absolute occurrence or non-occurrence, oblige us to think in terms of quantitative differences rather than gross qualitative ones.

If these points were always considered, the present bandwagon would probably roll rather more slowly, and would generate a greater percentage of interesting and reliable data.

4.3 THE HUNT FOR THE SIGNAL RECEPTOR

Nevertheless, there is no need to let the caveats become cavils. Nuclear location signals exist. They have been unequivocally demonstrated and characterized in several proteins. The implication of Bonner's studies mentioned at the start of this chapter, that nuclear proteins have 'homing devices', has been amply borne out. But of course, the very success of this enterprise has raised further questions, of which the most obvious is: if nuclear proteins have 'tickets' that allow them to enter the nucleus, what and where are the 'ticket inspectors'? There must be a signal receptor, or more likely a family of signal receptors, and the data discussed so far in this chapter make it likely but not certain that these receptors reside in the PCs. (The author first heard the 'ticket/inspector' analogy from Bob Lanford, and has heard several other workers in the field use it since.)

4.3.1 The hunt is difficult because the quarry is elusive

Newmeyer and his colleagues (Finlay *et al.*, 1987) used their *in vitro* system (the one using isolated nuclei resealed with oocyte cytoplasm: see Chapter 2) to demonstrate that 'ticket inspectors' really exist. Using human serum albumin cross-linked to the SV_{40} large-T nuclear location signal or a defective variant thereof (lysine-128 replaced by threonine), they obtained results very like those that Lanford and his colleagues obtained *in vivo*: proteins conjugated with wild-type but not those with defective signals entered the resealed nuclei quickly. The extent of uptake into the nuclei was proportional to the *in vivo* signal efficiency. This is a convincing vindication of their claim that their *in vitro* system behaves physiologically. Conjugates with defective signals did not compete with those with wild-type signals, but different proteins with wild-type signals did compete with one another, showing that the binding was saturable. The most striking result was that wild-type conjugates competed with nucleoplasmin. This was not a steric effect, because it was obtained with non-saturating concentrations of protein. The only conclusion seems to be that nucleoplasmin and the signal-linked protein compete for one receptor. Also, wheat germ agglutinin blocked translocation (see below), but it did not compete for binding with either nucleoplasmin or the wild-type conjugates, nor did these proteins have any effect on binding of the lectin. These results suggest two important conclusions. First, the same receptor recognizes different signals on different proteins; perhaps there is only one 'ticket inspector' after all. Secondly, receptor binding precedes translocation, and since translocation is blocked by wheat germ agglutinin it probably involves one or more of the PC glycoproteins. This supports the accepted view that nuclear location signals do indeed act at the NE; it is the 'missing step in the argument' (Section 4.2.2).

So there is strong experimental support for the prediction that at least one signal receptor exists. However, it has proved very elusive. Several laboratories began the hunt for the 'ticket inspector' in 1985–6, and only in 1989 have some strong candidates for the role been identified. Why has the task been so difficult? There seem to be two reasons. First, the receptor apparently has a low affinity for its ligands: Newmeyer gives an estimate of $10^{-7}M$ for the dissociation constant of the receptor–nucleoplasmin complex, and although the complexity of his system might throw doubt on this value (e.g. the activity coefficient of the protein in the oocyte cytoplasm is unknown), the value has been corroborated by other approaches (page 90). Secondly, all the ligands used, whether carrier-linked signal peptides or native proteins, show a lot of non-specific binding to nuclei. A low-abundance, low-affinity specific receptor is very difficult to find when it hides in a forest of nonspecific binding sites. Moreover, the poor primary sequence homology between different location signals handicaps experiment design somewhat: on which signal(s) should the investigator focus?

During the often frustrating hunt for the elusive 'inspector', a number of interesting observations have been made but their interpretation has been difficult. Goldfarb and his colleagues immobilized the isolated wild-type SV_{40} large-T location signal peptide and used this affinity chromatography medium to isolate salt-resistant signal-binding proteins from both total cytoplasmic and nuclear extracts. From the cytoplasm they obtained a 54 kDa protein. From the nucleus they obtained both this 54 kDa protein, and a 38 kDa protein which cross-reacted with monoclonal antibodies against Busch's nucleolar structural protein B23 – one of the components of the nucleolar part of the NS (Chapter 3). These results (information from David Goldfarb) await confirmation but they are interesting, because if the 54 and 38 kDa proteins were 'ticket inspectors', then the receptor isn't confined to the NE – it is distributed in the nuclear and cytoplasmic compartments. Also, Lanford and his colleagues prepared antibodies against the binding parts of the location signal, and cross-linked them to the signal peptide. They also succeeded in preparing and labelling anti-antibodies specific for the binding domain of the first antibody. By applying this labelled second antibody to cultured cells after the first, they hoped to label the signal-binding protein *in situ*. The same antibodies were used on Western blots of subcellular components, such as NE and its subfractions. The results pointed to a 56 kDa candidate, which the *in situ* experiments showed to be associated with an entire cytoplasmic network.

These studies suggest that the 'inspector', a 54–56 kDa polypeptide, is omnipresent in a cytoplasmic (and nuclear?) fibrillar system as well as the NE. The implications could be interesting. On the one hand, the findings raise the question of whether the solid-state model is to some extent applicable to protein transport as well as to mRNA transport (Chapter 1). On the other hand, they recall Maul's idea that PCs are transient entities, representing a 'snapshot' in the plane of the NE of a dynamic intercompartmental fibrillar system. These are attractive concepts. However, it now seems clear that Lanford's antibody labelling was not specific; and certainly the crucial experiment, showing that antibodies specific for the candidate inhibit protein transport *in situ*, was not performed successfully during these studies (information from Bob Lanford).

Uchida's group have performed just such a critical experiment, but their basic approach was quite different. Assuming that because the location signal is so basic the receptor must be acidic, they prepared antibodies against the sequence asp–asp–glu–asp. These antibodies gave the punctate immunofluorescence pattern on the nuclear periphery that is characteristic of PC-directed antibodies, and they inhibited the nuclear uptake of karyophilic proteins (Yoneda *et al.*, 1988). Could it be that the 'ticket inspector' has been partially characterized before it has been unequivocally identified?

4.3.2 Has the quarry been found?

The hunt seems at last to have been successful. So far, likely candidates for the 'ticket inspector' role have been identified in three laboratories: those of Gerace, Lanford and Riedel.

Lanford's and Riedel's groups (Yamasaki *et al.*, 1989; Benditt *et al.*, 1989) have taken similar approaches. Lanford's group modified a range of location signal peptides by incorporating a UV-photoactivatable cross-linker and also radio-iodinated them. Then they applied these covalent labels to whole cellular lysates. Four proteins (140, 100, 70 and 55 kDa) were labelled with relative intensities somewhat dependent on the actual location signal used. Labelling was blocked by unlabelled signal peptide but not by wheat germ agglutinin, and in partial confirmation of this team's earlier results (see above) these four proteins were not tightly PC-associated and showed a cytoplasmic and nuclear distribution. The results suggest that diverse location signals use a common transport pathway.

Riedel's group restricted their attention to the large-T signal. They bound the carrier-protein-linked signal to isolated NEs fixed onto glass slides and visualized the receptor–ligand complexes by indirect immunofluorescence. Complex formation was subject to competition by unconjugated wild-type peptide, but not by defective peptide (lacking lysine–128) or wheat-germ agglutinin. Conjugates of the defective peptide showed only minimal binding. Extraction with non-ionic detergent, which removes at least the ONM if not the INM (Chapter 2), virtually abolished binding, corroborating Lanford's inference of a loose association of the 'ticket inspector' with the PC, indeed with the NE as a whole. The results are also consistent with the inference from Newmeyer's work (see above) that receptor binding occurs outside the nucleus, before interaction with the PC, and does not involve the glycoproteins. When the wild-type signal peptide was modified by incorporation of ^{125}I and of an azido group to generate a photoaffinity label, four polypeptides with molecular weights 76, 67, 59 and 58 kDa became radiolabelled and these were extractable with non-ionic detergents. Two of these molecular weights are compatible with two of Lanford's (70 and 55 kDa). Benditt *et al.* may have observed the 76 kDa and the doublet at 58–59 kDa because of more efficient labelling or better resolved gels, or because of more proteolysis. They may have overlooked the 140 and 100 kDa proteins of Yamasaki *et al.* because of differential binding of the large-T and other signals. The work of these teams has also confirmed Newmeyer's value for the affinity constant of signal for receptor.

Gerace's group applied labelled signal peptides to the NE and cross-linked them to NE proteins using dimethyl suberimidate. The labelling was again subject to competition by unlabelled signal peptide. The labelled receptors were extracted with the nonionic detergent octyl glucoside (Adam *et al.*, 1989). Two receptor candidates were identified with SDS molecular weights of around 60 and 70 kDa, probably identical with the two largest of Riedel's polypeptides; but the identity cannot be confirmed without antibody evidence. Curiously, Adam *et al.* described the affinity of the receptor for the signal as 'high', but they quoted a dissociation constant (50–100 nM) which again accords very well with Newmeyer's value for nucleoplasmin.

4.4 PROTEIN TRANSLOCATION

4.4.1 The ATP-requiring step

Nucleocytoplasmic movement of a protein is at least a two-stage event. First, the signal binds to the receptor; this event seems to be rapid and energy-independent. Secondly, translocation occurs; this step is slower and apparently energy-dependent (Newmeyer and Forbes, 1988; Richardson *et al.*, 1988). The 'ticket inspector' performs his task quickly, but the 'turnstile' by which the protein enters the nucleus operates slowly and requires energy expenditure. The work that led up to these conclusions was more or less as follows (Figure 4.6).

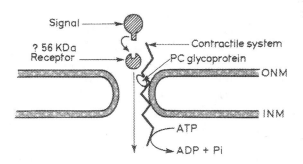

Figure 4.6 Schematic representation of specific nuclear protein uptake. The basic scheme is of the PC with a contractile apparatus, as shown in Figure 3.10 (the carbohydrate moieties of the PC glycoproteins have been omitted for clarity). The receptor molecule is associated with the contractile system which is continuous with intranuclear and cytoplasmic fibrils. It binds the signal on any signal-bearing ligand, which is then translocated with the aid of at least one of the PC glycoproteins and of ATP hydrolysis.

Newmeyer *et al.* (1986) showed that nucleoplasmin uptake into isolated resealed rat liver nuclei requires ATP. The addition of ATP did not alter the physiologically restrictive behaviour of the nuclei, which continued to exclude non-nuclear proteins. Non-hydrolysable analogues of ATP were not effective; and the temperature dependence of the uptake process closely simulated that found *in vivo* by Dingwall *et al.* (1982): nucleoplasmin is excluded at 0°C, equilibrates across the NE at 10°C and accumulates at 20–30°C. The only discrepancy was at 35–40°C, at which temperature Newmeyer's nuclei disintegrated, as all isolated nuclear preparations tend to do.

Although ATP depletion of the resealed nuclear preparations inhibited uptake of the protein, it did not inhibit its binding to the NE. Binding of the signal to the receptor is therefore independent of the functioning of the contractile or other translocating machinery. Whether the result also means that the signal receptor cannot be modified by phosphorylation is not yet clear; unambiguous identification of the receptor, and studies on isolated NE preparations, will be needed to settle this point.

Similar results have been obtained from *in situ* studies with Vero cells and *Xenopus* oocytes (Richardson *et al.*, 1988). Nucleoplasmin labelled with colloidal gold was microinjected and quickly became bound to PC-associated fibrils. Mutant nucleoplasmin lacking the signal sequence showed no such association. Cooling to 0°C or treatment of the cell with metabolic inhibitors prevented uptake through the PC into the nucleus but did not markedly alter the binding pattern. Results from experiments such as this are difficult to interpret because metabolic inhibitors tend to turn living cells into dead ones, so their effects on translocation could have been indirect; but the agreement with Newmeyer's *in vitro* findings is reassuring.

The ATPase involved in protein translocation has not so far been identified. Is it the NTPase? (Chapter 3). Quercetin, a potent inhibitor of the NTPase, seems to have no effect on nucleoplasmin uptake (information from Don Newmeyer), but despite the precautions taken in these experiments, the nature of Newmeyer's incubation medium makes the actual concentration of quercetin available to the NE uncertain.

At first sight, it seems curious that although nucleus : cytoplasm distributions of karyophilic proteins are maintained by binding and exclusion effects, not by active transport (Section 4.1), translocation through the PCs is nevertheless an active transport mechanism. But there need be no real contradiction here. The purpose of ATP hydrolysis in this system is surely to accelerate a process which in the absence of ATP would take several cell cycles; which of course would be lethally slow.

4.4.2 The effect of wheat germ agglutinin

Wheat germ agglutinin also inhibits nucleoplasmin uptake but does not block signal–receptor binding (see above). The effect of the lectin can be reversed by adding excess N-actylglucosamine. This suggests that although none of the PC glycoproteins is involved in signal receptor function, one of them is involved in translocation. The wheat germ agglutinin effect is not a simple steric block of the PC channels: the lectin has no effect on the passive permeation of fluorescently-labelled dextrans with molecular weights in the 10–20 kDa range. (This simple control is a prerequisite for interpreting data of this kind.)

Which glycoprotein is involved? Newmeyer's results suggested the 62–63 kDa protein (their molecular weight estimate was slightly higher, 63–65 kDa, but there is little doubt about the identity) which was described almost simultaneously by Davis and Blobel (1986). At the wheat germ agglutinin concentration they used (0.1 mg/ml) this was the only protein which bound a significant amount of the lectin. It is just possible that one of the other glycoproteins is involved and that this result is misleading; but if the identification is correct – which it is generally taken to be – then it has very interesting implications for the PC translocation machinery in general. Is Gp62 intimately associated with the ATP-hydrolysing and putative contractile machinery of PC?

4.5 OVERVIEW

Karyophilic proteins have two conjoint and distinctive properties: they are capable of entering the nucleus via the PCs, and they are capable of accumulating within the nuclear compartment. Accumulation depends largely on intranuclear binding, which is generally not well characterized, but cytoplasmic exclusion might play some part as well. Nucleocytoplasmic protein distributions are mainly established during interphase rather than during mitosis, when diffusible contents are excluded from the nuclear compartment.

Entry through the PCs depends on nuclear location signals, which are oligopeptide sequences within the mature protein molecules. There seem to be two broad classes of such signals: the paradigm cases are those in SV_{40} large-T antigen and in yeast MAT-α_2. The process of PC permeation involves at least two steps: binding of the signal sequence to a receptor, and translocation *per se*. So far, attempts to identify the signal receptor have produced several candidates (140, 100, 76, 67–70, 59–60 and 54–58 kDa polypeptides) which seem to be associated with cytoplasmic and perhaps nuclear fibrils as well as (loosely) with the NE. The receptor does not bind to wheat germ agglutinin, and is therefore not one of the PC glycoproteins identified by Gerace and his colleagues; it is detergent-extractable from NE, which suggests a membrane (probably ONM) location on the nuclear surface or very loose attachment to the PCs. Its affinity for ligands is rather low, around 10^7 M, suggesting that only very abundant nuclear proteins such as nucleoplasmin will saturate it fully *in vivo*.

Translocation *per se* is not yet well understood, but it seems to involve at least one of the PC glycoproteins, most probably the 62–63 kDa, and it may also involve the putative contractile apparatus of the PC. It certainly seems to depend on ATP hydrolysis. The enzyme involved in this has not been identified; despite some evidence to the contrary, the possibility that it is the NTPase remains open.

Readers who are interested in the development of this field but lack the time or inclination to pursue all the original references cited here can get alternative viewpoints by reading four of the reviews published over the last 5–6 years. In chronological order, these are: De Robertis (1983), Peters (1986: parts of this long article are about protein transport), Dingwall and Laskey (1986) and Paine (1988).

5 RNA transport

5.1 COUNTERFACTUALS AND ANALOGIES

This chapter begins by imagining a simple scenario: suppose we had all our present understanding of nucleocytoplasmic protein distributions, as discussed in Chapter 4, but we knew nothing at all about nucleocytoplasmic RNA transport. (In fact, we know quite a lot; our knowledge of RNA transport has grown in parallel with, and independently of, our knowledge of protein transport, and in this chapter the current picture is surveyed, but to begin with, just suppose that the picture were completely blank.) Given the scenario, can we use what we know about proteins as a basis for hypotheses about RNA transport? Presumably there is some commonality in the translocation machineries, because both classes of macromolecules use the same PCs, which can't be of unlimited complexity; but the obvious contrasts between the protein and RNA cases seem at first sight to exclude any useful comparison:

1. Proteins are imported from the cytoplasm; RNAs are mainly exported from the nucleus.
2. The cytoplasm does not generally seem to retain pools of karyophilic proteins (except in oocytes and perhaps in other cells in unusual metabolic situations), but most of the cell's RNA remains located in the nucleus.
3. Karyophilic proteins do not seem to need post-translational modification in order to enter the nucleus, but RNAs are usually made as immature precursors and undergo various kinds of post-transcriptional processing as prerequisites for export.
4. With a few exceptions, proteins seem to migrate alone from cytoplasm to nucleus, not associated with other proteins. Large RNA molecules, in

contrast, are not naked; they are always associated with specific sets of proteins.
5. At least some RNAs (messengers) move by a solid-state system and proteins do not. This seeming contrast might have to be re-evaluated if the protein signal receptors really are associated with extensive fibrillar networks in cytoplasm and nucleus.

But what does this list of contrasts really tell us? It implies what everyone knows: proteins and RNAs are entirely different sorts of molecules that do entirely different things inside the cell. But there is actually nothing in it to make us reject out of hand the idea that the mechanisms of nucleocytoplasmic transport could be similar for RNA and proteins. If we postulated such a similarity, merely as a source of working hypotheses, we could propose that:

1. An RNA molecule exported to the cytoplasm would have an intramolecular signal – some part of the sequence of the mature polymer which is essential for translocation.
2. The NE would contain one or more receptors for the signal(s). These receptors might be of low affinity, just as they are for proteins.
3. Binding of the RNA to the signal receptor in the envelope would be rapid, ATP-independent and insensitive to wheat germ agglutinin.
4. The actual translocation process would be slower, would require ATP hydrolysis, and would involve one of the PC glycoproteins (Gp62 if the analogy with protein transport is really close). This last characteristic would make translocation, as opposed to binding, susceptible to inhibition by wheat germ agglutinin.
5. The final distribution of RNA between the compartments might depend on binding within cytoplasm or nucleoplasm, though no doubt the rates of ancillary processes such as post-transcriptional modifications could predominate kinetically over the effects of binding sites.
6. Both translocation and intercompartmental binding might be regulatable, by modification of either the translocating or binding molecules or of the RNA itself.

This little discussion illustrates three important aspects of scientific thinking. First, we often use counterfactuals such as the opening scenario in exploring new areas of knowledge: arguments of the kind 'suppose the world were like such-and-such (though we know it isn't really), what could we conclude?' Philosophers of science have long been embroiled in controversies about the role of counterfactuals in science. From some philosophical perspectives, perhaps we shouldn't employ them; but in practice we do, because they're useful. They help us to get clear pictures of our own states of understanding. Secondly, science always makes use of analogies, which can be based on, amongst other things, counterfactual arguments. There is a right way and a wrong way to use analogy. If we took the hypotheses (1)–(6) above as credible descriptions of RNA transport, we'd be using analogy-as-justification, which is the wrong use. It could delude us into believing that we understand something that we don't understand at all. But if we were to use the same hypotheses merely as the targets for critical experiments, this would be analogy-as-heuristic, which in principle at least would be good science because it would open the door to the production of organized and informative experimental data. Thirdly, we can sometimes fail to see a potentially (heuristically) productive analogy because the two areas that we might compare seem superficially incomparable. Only when we examine this intuition of incomparability with due care and attention, as we just did for the case of protein versus RNA transport, can we overcome this barrier to potentially useful ideas. One of the best examples of this in the history of science is Clerk Maxwell's achievement in seeing an analogy, and ultimately a unification, between optics and electromagnetism.

This philosophical aside might seem inappropriate, especially since the scenario assuming ignorance of RNA transport is itself counterfactual; but one of the most surprising and most exciting conclusions to be drawn from recent work on nucleocytoplasmic mRNA transport, at least, is that the hypotheses based on our analogy with protein transport have all been corroborated. It is too early to say whether the same applies to other RNA classes. There has been little or no effective communication between researchers in the protein transport and RNA transport fields, and the above analogy-as-heuristic has not actually been applied. The experimental approaches taken to the study of mRNA transport have been radically different from those taken to the study of protein transport, and details of its mechanisms have emerged in a very different order. Therefore, the reassuring fact that our overall picture of mRNA transport now looks so similar to that of protein transport has largely escaped notice. Rational approaches to scientific inquiry of the kind outlined in this section are all very well, but real science often reveals a far more haphazard, far less rational course of historical development.

5.2 mRNA TRANSPORT

There is more discussion about the transport of mRNA than about the transport of tRNA, ribosomes or snRNPs in this chapter, mainly because we know more about it.

In Chapter 2, the study of nucleocytoplasmic transport by *in vitro* methods using isolated nuclei or resealed nuclear envelope vesicles was discussed. The point was made that several well-controlled studies during the last decade have demonstrated that mRNA efflux (the *in vitro* equivalent of transport) can usefully be studied using isolated nuclei, provided that nuclear swelling and degradative processes are prevented. There is some contamination with immature precursors in the lowest-abundance messenger class, but otherwise the system seems to behave physiologically normally. When resealed nuclear envelope vesicles containing trapped mRNA are studied *in vitro*, they show some leakage but the results can still illuminate the mysteries of the translocation machinery in the NE. Chapter 1 introduced the solid-state model of mRNA transport, according to which the *in vivo* transport process consists minimally of three steps: **release** from intranuclear binding sites, **translocation** through the NE and **cytoskeletal binding**. Efflux from isolated nuclei therefore represents – ideally – release plus translocation, while efflux from resealed NE vesicles represents – ideally – translocation alone. If the nuclear reticulum preparation (Chapter 2) really is equivalent to the *in situ* NS in respect of its RNA binding sites, then the release stage of mRNA transport can be studied using this preparation.

5.2.1 Post-transcriptional processing and RNP complexes

Eukaryotic mRNA is the product of post-transcriptional processing of a primary transcript. The main steps in processing (Figure 5.1) are: (i) capping, which is the transfer of several methyl groups to the $5'G$ end of the molecule; (ii) adenylation, which occurs in many but not all messengers, and is the addition of 150–200 adenylyl residues to the $3'$ end; and (iii) splicing, which is the removal of introns (intervening sequences) and the splicing together of the ends of successive exons. Capping occurs very rapidly, usually during transcription, and may play a part in the formation of initiation complexes during translation and in protecting the messenger against exoribonucleases. It may also be necessary for efficient splicing. Adenylation is an important contributor to the *in vivo* longevity of many eukaryotic messengers, and also seems to play a part in nucleocytoplasmic transport (see below). Splicing is the slowest step, and involves large complexes known as **spliceosomes**. Differential splicing can result in the production of several variant proteins from a single gene, as in the fibronectins and calcitonins. If processing is one of the stages in gene expression at which the cell's metabolism is controlled, then splicing is likely to be the regulatable step, though the control of mRNA half-life by altering the length of the poly(A) tail through the balance of poly(A) anabolic and catabolic enzymes might also be important (Figure 5.2). (See Müller *et al.*, 1985 and Agutter, 1988 for more detailed discussion.)

Messenger precursors in the nucleus (HnRNAs) are complexed with many proteins, of which six seem to form a 'core'. These six, known as A_1, A_2, B_1, B_2, C_1 and C_2 have molecular weights ranging from around 31 to around 43 kDa (Pederson, 1983). None of them is associated with mature mRNAs in

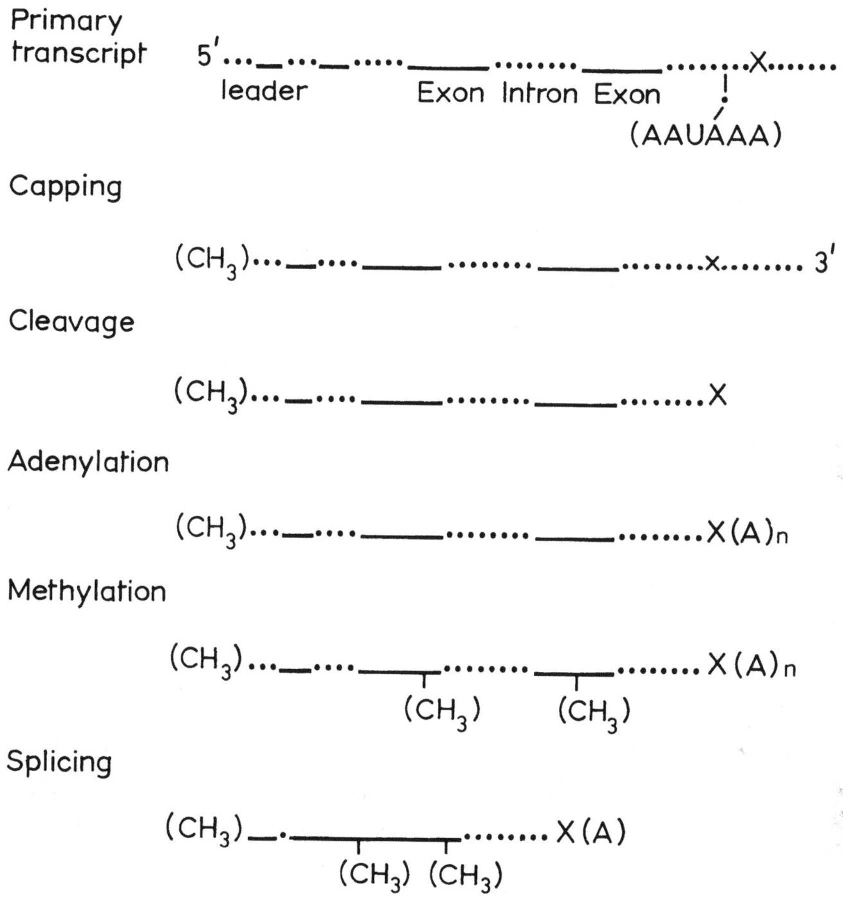

Primary
transcript   5′...__...__....._____........._____.........X.......
              leader   Exon Intron Exon   ┊
                                             ╱
                                         (AAUAAA)

Capping

(CH₃)...__...._____........._____.........x........ 3′

Cleavage

(CH₃)...__...._____........._____........X

Adenylation

(CH₃)...__...._____........._____........X(A)n

Methylation

(CH₃)...__...._____........._____........X(A)n
              ╷   ╷
            (CH₃)   (CH₃)

Splicing

(CH₃)__._____........X(A)
              ╷   ╷
            (CH₃) (CH₃)

Figure 5.1 Flow diagram summarizing the main steps in post-transcriptional processing of messengers.

the cytoplasm. They form a complex around which the RNA is wrapped, much as the DNA is wrapped round the core histones in a nucleosome. Conversely, at least one of the main cytoplasmic mRNA-associated proteins, a 75 kDa poly(A) binding protein first described by Blobel (1973), does not occur in the nucleus (van Eekelen *et al.*, 1981). It follows that at or near the translocation step, the mRNA is detached from its nuclear proteins and subsequently binds to its cytoplasmic ones. Electron microscope tomographic studies of the giant 35S messenger transcribed from the lampbrush chromosomes of *Chironomus* salivary glands (Skoglund *et al.*, 1983) show the

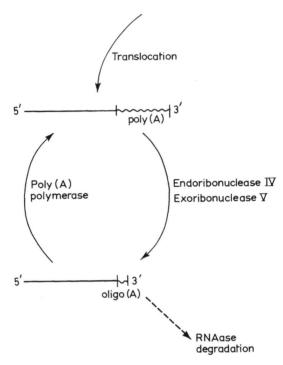

Figure 5.2   How poly(A) length is controlled. The balance between anabolic and catabolic enzymes, of which the main ones are named on the diagram, changes with metabolic circumstances and during ageing (Müller *et al.*, 1985). Shortening of the poly(A) tail is one way in which messengers are made more susceptible to degradation by RNAases.

intranuclear RNP particle as a tightly-coiled structure which uncoils as it reaches and touches the PC. After translocation the messenger, complexed with new proteins, coils up again. The first part of the RNA to contact the PC on the nucleoplasmic side is the 5′ end, but it could be that the messenger is translocated 3′ end first. (This messenger's translation products are the insect's salivary glycoproteins.)

5.2.2 Release

The main questions to answer about the release stage of mRNA transport are: (a) what does the HnRNA bind to in the nucleus? (b) What does the mature mRNA bind to in the nucleus? (c) How can we explain from the answers to (a) and (b) the fact that mRNA but not immature HnRNA gets released? (d) Can the process be controlled, and if so, how?

Some of the most distinguished contributors to our knowledge of the splicing mechanism have suggested that pre-messengers (splicing inter-mediates) are retained in the nuclear compartment because of the large size of the spliceosomes (Steitz, 1988). Spliceosomes (Fig. 5.3) comprise the pre-messenger (complete with at least one intron), several U-snRNPs and a collection of proteins that have yet to be adequately characterized. These particles are certainly fairly big, but there are several serious objections to this explanation of nuclear restriction. First, messengers themselves are

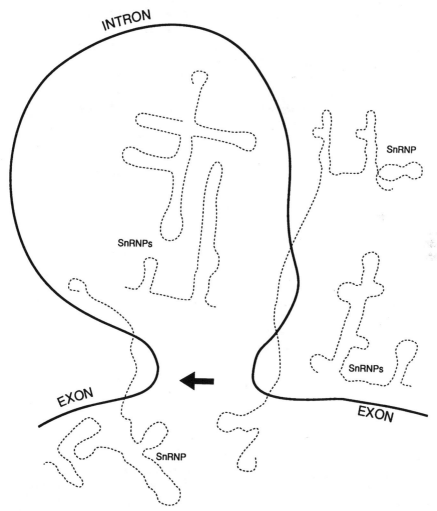

Figure 5.3 Diagrammatic model of a spliceosome, redrawn from Steitz (1988).

too large to enter the cytoplasm by diffusion, and all our knowledge of protein and RNA transport tells us that regulated nucleocytoplasmic exchange processes are not diffusion processes. Secondly, particles larger than spliceosomes – colloidal gold particles, for example – pass through the PCs rapidly *in situ* if coated with appropriate signal-bearing materials. Thirdly, isolated nuclei under suitable conditions (Chapter 2) exhibit normal restriction of pre-messengers, despite their indubitable leakiness which would ensure the rapid loss of non-immobilized spliceosomes. Fourthly, there is no evidence that NE puncturing allows spliceosomes to leak into the cytoplasm. Fifthly, in situations where immature mRNA precursors do get into the cytoplasm (e.g. in some cases of carcinogenesis), there is no evidence that spliceosomes are abnormally small or unstable.

The most important study of the release stage of mRNA transport to date was carried out by Schröder *et al.* (1987). Using quail oviducts, they found that both the ovalbumin messenger and its well-characterized splicing intermediates were recovered in the nuclear reticulum isolated by the Comerford *et al.* (1986) procedure. The splicing intermediates were more or

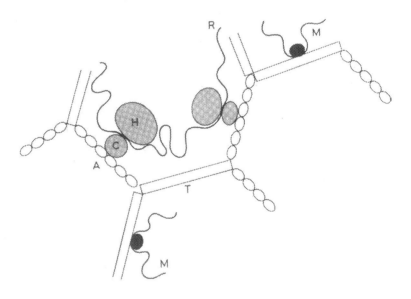

Figure 5.4 A modified version of the model of HnRNP/NS attachment shown in Figure 3.12, incorporating the idea that mature messengers are associated with the DNA topoisomerase II regions of the NS rather than the actin regions (Schröder *et al.*, 1987). M = messenger RNA; other symbols as in Figure 3.12.

less wholly recovered; some of the mature messenger was lost during isolation of the reticulum, suggesting that its binding might be more labile. When the preparation was treated with F-actin disrupting agents such as cytochalasin B, the splicing intermediates were selectively solubilized; the mature messenger was retained in the pellet fraction. In contrast, low concentrations (around 10 μM) of ATP or a non-hydrolysable analogue of ATP eluted the mature messenger, not the precursors, and this elution was prevented by inhibitors of DNA topoisomerase II. If the reticulum is assumed to represent the NS, these results suggest that immature messenger precursors are attached (perhaps via the introns) to actin-containing parts of the NS, and the mature mRNA in the nucleus is attached to DNA topoisomerase II-containing parts (Fig. 5.4).

Together with the evidence that HnRNP is attached to the nuclear matrix via the core C-group proteins (van Eekelen and van Venrooij, 1981), which are associated with exons (Pederson, 1983), this suggests an interaction between the C-group proteins and actin. The results from Müller's team (Schröder *et al.*, 1987) might also throw light on the curious observation that intranuclear microinjection of an isolated gene leads to normal transcription, processing and transport of the messenger to the cytoplasm if and only if at least one intron is retained in the gene (if all the introns are removed before microinjection, there is normal transcription but the messenger remains intractably bound in the nucleus; Gruss *et al.*, 1979). Presumably the transfer from the actin binding site to the DNA topoisomerase II binding site is an essential step during release. Further to this point: if mRNA transport is selective – that is, if only some of the stock of mature messengers in the nucleus is exported to the cytoplasm at any one time – then the selectivity presumably hinges on the dynamics of DNA topoisomerase II. Knowledge of how this enzyme is controlled could generate useful hypotheses about the selectivity of transport, and about the breakdown of transport in early carcinogenesis. It is interesting that in some systems at least, DNA topoisomerase II can be phosphorylated by intranuclear protein kinase C (Rottmann *et al.*, 1987), which is of course activated by co-carcinogens such as phorbol esters. Note that an assumption of this underlying argument is that the NS is not only real but also includes actin and DNA topoisomerase II among its core polypeptides.

Experiments in Zasloff's laboratory have shown that messengers micro-injected into *Xenopus* oocyte nuclei are efficiently exported to the cytoplasm only when they contain promoter region transcripts (De la Peña and Zasloff, 1987). It is of course dangerous to generalize from data obtained with this type of cell, but a possible explanation for these rather odd results is that promoter transcripts are necessary for the kind of attachment to the NS that makes subsequent release of messengers possible.

5.2.3 Translocation

Around 1970, studies by Ishikawa and his colleagues suggested that mRNA efflux from isolated liver nuclei was stimulated by ATP. These *in vitro* systems were ill-characterized and it was not clear that the RNA in the post-incubation supernatants was mainly, let alone exclusively, mature messenger. However, with the advent of better-characterized systems, the ATP dependence of mRNA efflux was established at least in the case of liver nuclei. By comparison of the properties of ATP-stimulated mRNA efflux with those of enzymes in isolated NE preparations, the probable role of the NTPase in mRNA translocation was established. The role of ATP hydrolysis was therefore the first thing to be established about messenger transport; it is almost the latest about protein transport. This early work has been reviewed several times (Agutter, 1988). Here, only a summary of the main lines of evidence for the NTPase/translocation relationship is given.

1. The kinetic properties of the enzyme and of mRNA efflux are closely comparable. Under the same conditions, both show the same apparent K_m for ATP; this parameter changes in the same way for both NTPase and efflux process as the conditions are changed; and both the enzyme's initial velocity and the rate of mRNA efflux show a similarly non-Michaelis–Menten dependence on ATP concentration under one set of conditions, and a Michaelis–Menten dependence under others.
2. NTPase inhibitors inhibit ATP-dependent mRNA efflux, generally showing the same half-maximal concentrations. Both the NTPase and the efflux process show the same substrate and activator specificity.

These data were first obtained in the author's laboratory (Agutter *et al.*, 1976) and were subsequently corroborated by others (Clawson *et al.*, 1980a; Bernd *et al.*, 1982).

3. Changes in the mRNA transport rate *in vivo*, e.g. those induced by carcinogens and by other drugs such as tryptophan, are paralleled by increases in the rate of mRNA efflux from isolated nuclei and by increases in the NTPase activity in isolated NEs (Clawson *et al.*, 1980b; Murty *et al.*, 1980).
4. Export of poly(A)+mRNA from resealed NE vesicles shows the same dependence on ATP hydrolysis as efflux from isolated nuclei (Riedel *et al.*, 1987; Agutter, 1990).
5. A monoclonal antibody that blocks mRNA efflux also inhibits the NTPase (Baglia and Maul, 1983). At the time of publication of this work, Baglia and Maul believed the monoclonal to be lamin B specific, but this was based on the evidence of Western blots of 1-dimensional SDS gels. It seems more likely that the epitope was on Gp62, which virtually co-migrates with lamin B in some systems; certainly wheat germ agglutinin

blocked the efflux process as well, though not the NTPase, a finding corroborated by more recent *in situ* studies (Davis and Blobel, 1986).

This weight of evidence seems compelling, but it has to be emphasized that in some cell types, e.g. some cultured hepatoma and myeloma lines (Schumm *et al.*, 1977; Otegui and Patterson, 1981), mRNA efflux (and therefore *in vivo* transport?) is ATP-independent. It should also be emphasized that the ATP dependence of translocation in tissues such as liver cannot be explained in terms of ATP stimulation of release from DNA topoisomerase II-related binding sites (Section 5.2.2). In translocation, non-hydrolysable analogues cannot substitute for ATP, and in any case the K_m value is very much higher (in the order of 100–200 μM rather than 1–10 μM).

The next point (chronologically) to be established about mRNA transport concerned the intramolecular signals in messengers and the signal receptor in the PC. A single class of saturable poly(A)+mRNA binding sites has been demonstrated in NEs, with a fair degree of specificity for poly(A). Any oligo(A) or poly(A) of more than about 15 residues has maximal affinity for these putative receptors. Not surprisingly, when poly(A) or messenger is bound to these sites, the NTPase is stimulated; these findings have again been independently corroborated (Agutter *et al.*, 1977; McDonald and Agutter, 1980; Bernd *et al.*, 1982). The dissociation constant is high, around 3×10^{-7} M, but is decreased by endogenous phosphorylation to around 10^{-7} M, still in the same order as the affinity of the protein 'ticket

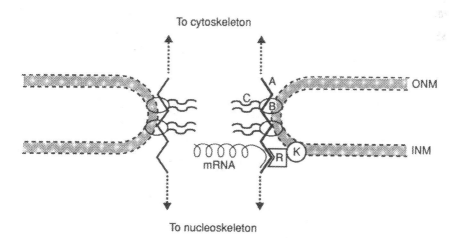

Figure 5.5 Control of the mRNA signal receptor (R) in the PC by phosphorylation via the endogenous protein kinase C (K), which appears to be closely associated with the receptor (Schröder *et al.*, 1988a; S. J. M. Aitken, personal communication). A = contractile apparatus, B = PC glycoprotein, as in Figures 3.10 and 4.6.

inspectors' for their 'tickets'. At least some of the endogenous phosphoryl-ation is attributable to protein kinase C (Figure 5.5), in which either the α or the β isoform is present in liver NEs (Schröder *et al.*, 1988a, corroborated by Stuart Aitken in my laboratory). The prospects for metabolic regulation of mRNA translocation are obvious, and these results could explain the parallel stimulation of mRNA transport and NTPase activity in liver by insulin (Purrello *et al.*, 1982).

So it seems that poly(A) is one intramolecular signal by virtue of which mRNAs can interact with the translocation machinery. Obviously it isn't the only such signal, because not all mRNAs are adenylated. Indeed, in the amphibian oocyte, the signalling system might be very non-specific; recent results from Feldherr's laboratory show that when colloidal gold is coated with polyribonucleotides and microinjected into the nuclei of these cells it is exported apparently irrespective of the base composition of the polymer used (Dworetzky and Feldherr, 1988). However, it is just possible that these particles cannot interact normally with the solid-state machinery: cf. the

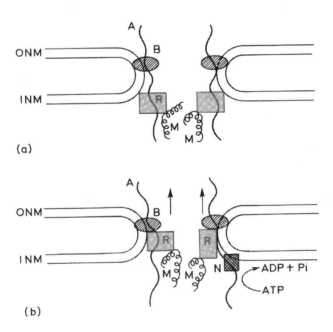

Figure 5.6 Two possible mechanisms for mRNA translocation; key as in Figure 5.1. The left-hand side of the annulus shows the mRNA (M) bound to the receptor being translocated in conjunction with a PC glycoprotein, such as Gp62. The right-hand side shows essentially the same thing, except that the receptor (R) has fragmented and one of its fragments, the NTPase (N), acts as the motor driving the translocation process.

discussion in Section 5.2.2. of the results from Müller's and Zasloff's laboratories.

What else do our *in vitro* results tell us? First, there is a signal receptor in the envelope, the affinity of which for its ligands is fairly low but can be modified by endogenous phosphorylation. Secondly, this phosphorylation is catalysed by an enzyme that is known to be sensitive to various hormones, growth factors and cancer promoters. And thirdly, in liver and other tissues, though not in cancer cell lines, mRNA translocation is dependent on ATP hydrolysis. It is also clear that the ATP is hydrolysed not when the poly(A)+ mRNA binds to the envelope, but subsequently, during translocation *per se* (McDonald and Agutter, 1980). Moreover, the translocation event can be blocked by wheat germ agglutinin and seems to involve Gp62 (Fig. 5.6). In all these respects the analogy between mRNA translocation and protein translocation seems remarkably close, and there are grounds for supposing the process to be pertinent to metabolic regulation.

The 'hunt for the ticket inspector' in the case of proteins has been frustratingly difficult; in the case of mRNA it was in one respect rather .easier. There is rather less non-specific binding of poly(A) than of proteins to the NE, particularly if the membranes are removed and a PCLF preparation is used (Chapter 2). Therefore, one approach to the problem, using signals modified by the introduction of covalent cross-linkers, can be deployed successfully. Another approach is to see which of the NE polypeptides is phosphorylated differently in the presence and the absence of poly(A), since poly(A) and some of its analogues inhibit phosphorylation and promote dephosphorylation (McDonald and Agutter, 1980; Bernd *et al.*, 1982). Both these approaches have been used and have given clear and reproducible results. However, the hunt has been easier in one respect only than it has in the case of proteins. In another respect it has been at least as difficult. The results given by the two experimental approaches were, or seemed, incompatible.

The covalent cross-linking approach was developed in Fasold's laboratory (Prochnow *et al.*, 1990). They used radiolabelled poly(A) with about 10% of the A residues replaced by 8-azido(A), giving a photoaffinity label. Interestingly, they showed that the signal receptor for poly(A)+mRNA can be labelled efficiently from the inside but not the outside of a resealed NE vesicle by this reagent, suggesting the kind of asymmetric receptor distribution that we might expect. On SDS gel electrophoresis and autoradiography they found a single labelled band of 270 kDa, which on treatment with RNAases or at high pH generated two polypeptides of around 50 and 30 kDa. The phosphorylation-inhibition approach was used in my laboratory and in Müller's (Agutter, 1985a; Schröder *et al.*, 1988a). We both identified a likely-looking polypeptide of 106–110 kDa (hereafter called P110). A monoclonal antibody against P110 blocks the export of poly(A)+

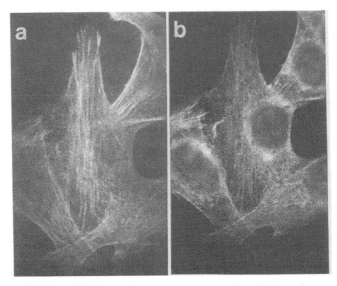

Figure 5.7 Immunofluorescence picture showing the cytoplasmic distribution of the mRNA signal receptor protein. Photograph kindly supplied by Dr M. Bachmann and Prof. Dr W. E. G. Müller, and reproduced by permission of the Editorial Board of *Arch. Biochem. Biophys.*

RNA from resealed NE vesicles, so long as it is entrapped in the vesicles not just added to the incubation buffer (Agutter, 1989).

This contradiction in molecular weights is not so fundamental as it first seemed. P110 is very proteinase-labile, and endogenous proteolysis in isolated NEs yields a family of polypeptide fragments, including two at about 50 and 30 kDa. These fragments also bind to poly(A). Detlef Prochnow and Stuart Aitken in my laboratory have recently found that the 270 kDa 8-azido-poly(A) labelled complex also breaks down to give apparently the same family of polypeptide products. Interestingly, one of these has the same molecular weight as the NTPase, which has been purified and characterized (Schröder *et al.*, 1986a), and has the same capacity to bind 8-azido-ATP. Is it possible that P110 (or its larger parent) is a myosin-like protein, cleavage of the ATP-hydrolysing fragment from which is stimulated by poly(A) (cf. Berrios and Fisher, 1986; Schindler and Jiang, 1986)? This would certainly explain the increase in NTPase activity in the presence of poly(A), and the diminution of the phosphorylated 110 kDa band in the same conditions. Unfortunately, our only monoclonal antibody against this receptor does not label all the putative fragments on Western blots. It reacts most strongly with the 66 and 58 kDa fragments, and more weakly with the 96 kDa and the 110 kDa itself. Whether the other bands contain blocked or incomplete epitope, or whether they are irrelevant polypeptides, is not yet completely clear.

In any case, so far as poly(A) and poly(A)+mRNA binding are concerned, we do seem to be dealing with only one protein. This is consistent with the linearity of the Scatchard plots of poly(A) binding to NEs, or at least to PCLFs (McDonald and Agutter, 1980). The only problem is to find out what its real molecular weight is. This can best be done by isolating and sequencing the gene, a project that we have recently started. It is possible that the 110 kDa PC-located protein recently described by Aris and Blobel (1989) is the same protein; it seems to have 95 and 55 kDa fragments. Aris and Blobel have identified a protein in yeasts that cross-react with an antibody to their 110 kDa polypeptide, and given the advanced state of yeast molecular genetics this might be a valuable step towards gene isolation and sequencing.

Protein transport from cytoplasm to nucleus has been studied almost exclusively by *in situ* methods, and while these experiments have given us excellent topological information we are still largely ignorant of the kinetics of the process. Messenger RNA translocation from nucleus to cytoplasm has been studied largely by *in vitro* methods, and while the results allow us to construct detailed kinetic models of the process (Agutter, 1988) they have failed to supply us with a topological model, in which we can relate the events in translocation to the molecular architecture of the PC. The two experimental approaches, *in situ* and *in vitro*, have complementary strengths and complementary weaknesses.

5.2.4 Cytoskeletal binding

Only one piece of information about cytoskeletal binding is well established: it involves actin. Translationally active mRNA – polysomal material – is all bound to the cytoplasmic microfilaments or to the two-dimensional 'protein gel' contiguous with the cytoplasmic face of the rough ER, a structure that also apparently contains actin fibrils. There are no such things as 'free polysomes'. The evidence for this includes localization of specific viral messengers, the effect of microfilament disrupting agents on translation, and the co-purification of translationally active messengers with the appropriate parts of the cytoskeleton. It was well and briefly reviewed recently (Jones and Kirkpatrick, 1988). Just how the mRNA is bound to the actin is not clear yet, but the evidence discussed in the following section might throw some light on it.

5.2.5 The 'cytosolic stimulating proteins'

A long series of studies by Dorothy Schumm and Thomas Webb and their colleagues has demonstrated that the 100 000 g supernatant of a tissue homogenate contains protein factors that stimulate mRNA efflux from isolated homologous nuclei (e.g. Palayoor *et al.*, 1981). The factors present in regenerating liver are more active than those in adult liver, and hepatomas contain very active (and possibly different) factors. The liver factors co-purify to a large extent with polysomes, from which they can be obtained by differential salt extraction and purified (Moffett and Webb, 1983), e.g. by ion-exchange and poly(A)-Sepharose chromatography. Two polypeptides purified in this way, molecular weights 58 and 31 kDa (Schröder *et al.*, 1986b: the 31 kDa probably corresponds to the 35 kDa protein isolated by Moffett and Webb), have an interesting set of characteristics: they bind poly(A), stimulate the NTPase in isolated nuclear envelopes, and also promote mRNA efflux. More rapid purification by HPLC reveals that the 58 and 31 kDa polypeptides are two members of a now familiar family: P110 and its degradation products. Moreover, the monoclonal antibody against P110 has been shown by immunofluorescence to react with a set of cytoskeletal fibres, at least some of which are microfilaments (Schröder *et al.*, 1988b; Figure 5.7). Incidentally, the best-known and most abundant poly(A) binding protein in the cytoplasm, the 75 kDa protein described by Blobel (1973), is not a member of the family. It is extracted from polysomes by much lower salt concentrations than the P110 series (Schröder *et al.*, 1986b), and it prevents rather than promotes poly(A) stimulation of the NTPase in isolated NEs (Bernd *et al.*, 1982).

The case of P110 thus resembles the case of the signal receptor for protein transport: P110 seems to be located on cytoplasmic fibrils as well as the PCs. Moreover, it might not be confined to the cytoplasm. Schweiger and Kostka (1984) have found a poly(A) binding protein of just the same molecular weight, and with just the same tendency to fragment into smaller polypeptides, also of the same molecular weights, in HnRNP particles. It seems likely that this protein is P110. If so, it presumably plays a part in attaching the HnRNP core to the actin filaments of the NS (cf. Schröder *et al.*, 1987), and could well do the same in the cytoplasm (Jones and Kirkpatrick, 1988; Schröder *et al.*, 1988b). If these speculations are valid, then P110 is an integral part of the solid-state transport machinery.

5.3 tRNA TRANSPORT

Our understanding of tRNA transport has come from *in situ* rather than *in vitro* studies, mainly the work of Zasloff and his colleagues. They performed site-directed mutagenesis on the human initiator-methionyl tRNA (tRNA$^i_{met}$) gene and injected the tritiated wild-type or mutant tRNA (they have obtained more than 30 mutants to date) into amphibian oocyte nuclei. In some experiments, they injected other (unlabelled) tRNAs at the same time. After microinjection, they examined the nucleocytoplasmic transport rate either by autoradiography or by microdissection and direct measurement of label in the main cell compartments. Their conclusions (Zasloff, 1983; Tobian *et al.*, 1984, 1985) are:

1. The tRNA transport mechanism shows saturation kinetics, suggesting that a facilitated transport process is involved. Zasloff and his colleagues suggested that the process is not active because ATP depletion of the cells did not impair it. This is scarcely sound evidence (see Chapter 4 for discussion of similar experiments on protein transport), but the conclusion seems valid (see below).
2. All tRNAs examined compete with each other for transport. This suggests a single transport system for all tRNAs. At least, it indicates that there is at least one common (rate-limiting) step in the mechanisms of transport of all tRNAs.
3. The 'K_m' for tRNA transport, at least for tRNA$^i_{met}$, is around 10^{-7} M.

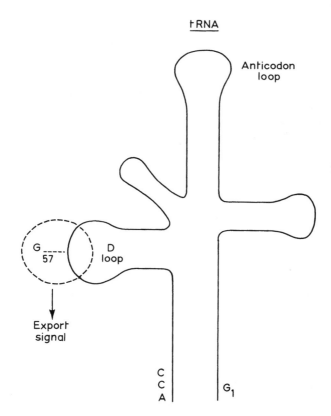

Figure 5.8 Structure of the tRNA molecule, showing the intramolecular signal around residue G_{57}: CCA = 3' terminal triplet, G = guanosine.

This value seems to be common to all PC signal receptors for all transportable macromolecules! In the case of tRNA, at least, it suggests that the system is probably not saturated under physiological conditions.

4. The export signal sequence(s) is/are probably in the highly-conserved D and T loops of the molecule. In particular, if G_{57} is replaced by another base, the transport rate is decreased around 20-fold (Fig. 5.8).

5. Post-transcriptional processing of the tRNA precursor is mechanistically quite distinct from transport. Mutations that impair transport do not impair removal of the single 8-base intron, or vice-versa.

6. Microinjected tRNA does not re-enter the nucleus from the cytoplasm at a significant rate.

These experiments have been well conceived and executed, but a possible objection to them is that a foreign tRNA injected in often rather large quantities into an oocyte nucleus might not behave physiologically. For example, if physiologically there is some kind of solid-state tRNA transport machinery, then the microinjected material might not become functionally incorporated into it. Another objection is the familiar one that the amphibian oocyte might be atypical in other respects than sheer size, and Zasloff's findings might not be universally valid. On the other hand, the analogy between what Zasloff *et al.* have found for tRNA transport and what we know about protein transport and mRNA translocation (specific intra-molecular signals; signal receptors – presumably in the PC – for which different signal-bearing molecules compete; similar receptor affinities for ligands) is once more reassuring.

Gerace's group have examined the effects of their monoclonal antibodies against PC glycoproteins on tRNA transport in amphibian oocytes. There have been some interpretation problems because of the antibodies' tendency to cross-react (Chapter 3), but comparative studies with different mono-clonals combined with the Western blotting evidence overcome these. The results suggest that antibodies reacting with Gp62, injected into the nucleus but not into the cytoplasm, inhibit tRNA export. For example, the antibody RL1, which reacts most strongly with Gp62, inhibits tRNA export most powerfully: around 4–5-fold (Featherstone *et al.*, 1988). Once again, a familiar pattern emerges: Gp62 is involved in tRNA translocation, just as it is in mRNA translocation and protein translocation. The same antibody that blocks the export of tRNA blocks the import of proteins.

Indeed, the general picture is so similar: specific intramolecular signals in the mature molecule, saturable signal receptors in the NE with affinities in the 10^7 M range, involvement of Gp62 and so on; that it's tempting to generalize further. Could tRNA transport involve a solid-state machinery as well, and could it, despite Zasloff's evidence to the contrary, require ATP hydrolysis at the NE? Marian Thomson in my laboratory has tested this last possibility using the resealed liver NE vesicle system. Saturable, unidirectional export of tRNA is observed as per prediction. However, not only does ATP fail to stimulate this export, ATP and other nucleoside triphosphates actually inhibit it. Also, the export of tRNA is not blocked by the monoclonal that blocks mRNA export. The analogies between the systems for protein, mRNA and tRNA are certainly close, but the commonality is not total.

5.4 RIBOSOME TRANSPORT

tRNA transport is almost as well understood as protein and mRNA transport; but this happy state of affairs does not extend to ribosomes and ribosomal RNA. The ribosomal proteins become associated with the rRNAs during processing in the nucleolus, and the two-ribosomal subunits are exported separately. There is a mutant of *Saccharomyces cerevisiae* in which large subunits accumulate in the cytoplasm; small subunits are made, but apparently remain in the nucleus (Carter *et al.*, 1980). This suggests that the mechanisms of transport of ribosomal subunits are distinct and probably rather interesting, but it only throws our present state of ignorance into relief.

Some attempts have been made to characterize ribosome transport *in vitro* using isolated rat liver nuclei. The results suggest that ribosomal subunit transport depends on ATP hydrolysis and on proteins present in the 100 000 g supernatant of the liver homogenate. These proteins do not seem to be ribosomal constituents *per se*, but to date they have not been characterized and their mode of action is unknown (Schumm *et al.*, 1979). Unfortunately, it is not entirely clear that the isolated nuclei behave physiologically with respect to ribosome transport, or that the ribosomal subunits found in the post-incubation supernatants are normal, or that at least some of the subunits have not simply been desorbed from the surfaces of the nuclei. Nevertheless, some preliminary experiments with resealed NE vesicles support the conclusions that cytosolic proteins are necessary for ribosomal transport, though the role of ATP has not been confirmed by these studies (information from Iris Hassell and Hugo Fasold).

At present, at least two experimental approaches are being taken to this topic: further *in vitro* studies using the resealed NE vesicles, and *in situ* studies in which rRNAs (and modified forms thereof) are linked to colloidal gold are microinjected into amphibian oocyte nuclei (information from Howard Fried). With any luck, these two approaches will give us the complementary benefits of *in vitro* and *in situ* studies.

5.5 snRNA TRANSPORT

Some very informative studies on the transport of UsnRNAs and snRNPs have been conducted by De Robertis and his colleagues. Once again, the amphibian oocyte provided the experimental material. The oocyte cytoplasm contains a pool of the protein components of snRNPs such as U_1 and U_2. These proteins are not themselves karyophilic: when they are isolated (e.g. from snRNPs) and microinjected into oocytes, they remain exclusively cytoplasmic. When U_1 or U_2snRNA is microinjected into the cytoplasm, it complexes with proteins from this pool to form a normal snRNP particle, and

SnRNP

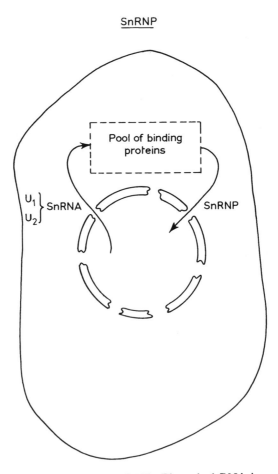

Figure 5.9 Shuttling of a UsnRNA/RNP. The naked RNA is exported from the nucleus, associates specifically with proteins from the cytoplasmic pool, and re-enters the nucleus as a complete RNP particle.

this particle rapidly enters the nucleus. When mutant oocytes with depleted or absent pools of these proteins are injected in the same way, the snRNA remains in the cytoplasm: it can enter the nucleus only when it is complexed with the appropriate proteins. UsnRNAs are, of course, made in the nucleus. These results imply that the naked snRNA is exported from the nucleus, binds to its proteins in the cytoplasm, and immediately re-enters the nucleus as a snRNP complex (Mattaj and De Robertis, 1985) (Figure 5.9). This makes sense of the long-known shuttling of proteins and small RNAs between nucleus and cytoplasm in some species (Jelinek and Goldstein, 1973).

What does this surprising information tell us about the signals involved in snRNP transport? Since neither the RNA alone nor the proteins alone are karyophilic but the complex is, there are three possibilities:

1. The signal is on one or more of the proteins, but is not exposed (not visible to the PC receptor) until the protein(s) in question is/are complexed to the RNA.
2. The signal is on the RNA, but is not visible to the receptor until the RNA is complexed to the protein(s).
3. The signal is not part of the primary structure of either the RNA nor any of the proteins, but is part of the quaternary structure of the complex.

The evidence to date favours possibility (2). Mutants of U_2snRNP have been studied by microinjection into the cytoplasm (Konings and Mattaj,

U_2 snRNA

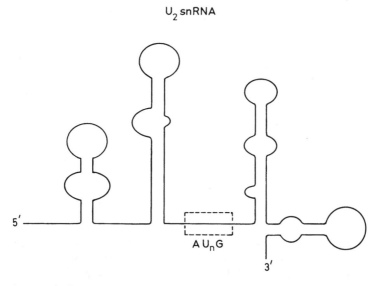

Figure 5.10 Schematic representation of the secondary structure of U_2snRNA. showing the proposed signal sequence AU_nG.

1987). There is a sequence, AU_nG, in the 3′ half of the RNA molecule which seems to be essential for uptake into the nucleus (n is about 6, but the exact number may not be crucial) (Figure. 5.10). Moreover, the same study indicated that this sequence – the translocation signal – is not exposed in the naked RNA but becomes exposed when the RNP complex forms. This presents the possibility of another means of controlling nucleocytoplasmic transport processes: altering the tertiary or quaternary structure of a macromolecule so as to expose, or not to expose, the translocation signal, thus altering the ability of the ligand to bind to the NE receptor.

5.6 OVERVIEW

In this chapter, it is emphasized that good *in situ* and good *in vitro* studies complement one another. The former give very reliable qualitative information that can be matched to the molecular architecture of the intracellular machinery, but generally cannot give quantitative (especially kinetic) information. For the latter, the reverse is true. The point is illustrated by comparing our knowledge of mRNA transport, especially the NE (translocation) step, with our knowledge of protein transport. The former is largely based on *in vitro* data, the latter on *in situ* data. Given this, it is reassuring, and perhaps surprising, that our emerging pictures of protein transport and mRNA transport are so similar. The main points of resemblance are:

1. There are signal sequences within each mature transportable macromolecule. Some of these have been characterized, at least provisionally.
2. There are receptors for these signals in the NE.
3. These receptors have a low affinity (around 10^7 M) for their ligands, but the affinities might be changed by modification of either ligand or receptor.
4. The receptors are not confined to the PC, but appear in cytoskeletal fibres and, probably, within the nucleus. This might be relevant to binding within nucleus or cytoplasm or both.
5. Binding of ligand to receptor is ATP-independent.
6. Translocation of the ligand is dependent on ATP hydrolysis.
7. Translocation of the ligand also depends on one of the PC glycoproteins, Gp62, and is consequently inhibited by wheat germ agglutinin.

All these points have been established for proteins (Chapter 4), and they have been established for mRNA. One of the signals in mRNA is poly(A), and its receptor is a labile protein which typically has an SDS molecular weight of about 110 kDa (P110). The affinity of this protein for mRNA can be increased by endogenous phosphorylation, e.g. by protein kinase C. P110 is present in cytoplasmic microfilaments, to which polysomal mRNA is known to be bound, and may be present in NS-bound HnRNP particles. Within the nucleus, mature mRNA seems to be bound to the DNA topoisomerase II parts of the NS, and its immature precursors to the actin parts; interestingly, DNA topoisomerase II might also be phosphorylated by protein kinase C. Messengers bind to P110 in an ATP-independent fashion, but translocation seems to depend on ATP hydrolysis by a broad-specificity 40–45 kDa enzyme, the NTPase, which might itself be a fragment of the receptor (P110) molecule. Finally, translocation (and the NTPase) can be blocked by binding wheat germ agglutinin to Gp62.

In the case of tRNA, the signal region (around G_{59}) is present in the D/T loops. The receptor has not been identified, so its distribution and modifiability are unknown, but it clearly exists: tRNA translocation is saturable and once again the affinity is around 10^7 M. ATP hydrolysis does not seem to be involved in tRNA translocation. Gp62 seems to be involved in translocation of tRNA as well as mRNA and proteins.

For snRNPs, our information is still less complete. There are certainly intramolecular signals, of which the AU_nG on U2snRNA is characterized. This signal can be modified in a very interesting way: its accessibility, and presumably its ability to bind with the (putative) receptor in the NE, depends on the appropriate interactions of the snRNA with its associated proteins. There is no convincing information about the nature of the receptor, the involvement of ATP hydrolysis or the role of Gp62 in the case of snRNPs.

For ribosomal subunits, our information is limited to the facts that the subunits are transported separately, they must be intact (rRNA alone is not efficiently exported from resealed NE vesicles), and the process is stimulated by one or more unidentified cytosolic proteins. There is a conflict of evidence about the importance of ATP hydrolysis in the process. No intramolecular signal in any ribosomal component has been identified, and there are no clues about the identity of the receptor or the involvement of Gp62. The 145 kDa nucleolar skeletal protein has been implicated in binding and transport of ribosomal subunits (Chapter 3) but there is no information about the nature of ribosome binding to this protein or about the release mechanism.

6 *The importance of nucleocytoplasmic transport*

So far, the data available in a single field of research, and the ways in which these data are obtained and interpreted, have been surveyed. This final chapter is different. 'Importance' is not a matter of data and interpretation; it is a value-judgement. Therefore, many of the arguments put forward in these remaining few pages have less claim to 'objectivity' (whatever is meant by that notoriously difficult word) than those in the previous five chapters, and they should be judged accordingly. The word 'importance' is used here in two main senses. First, it means the extent to which recently-acquired understanding of nucleocytoplasmic transport processes contributes to an understanding of biology in general and cell biology in particular. Secondly, less obviously, it means the extent to which this piece of recent scientific history throws light on the way in which modern research operates: can any inferences be drawn about the nature of science (or at least of cell and molecular biology) from a survey of this one specialized field? At various points in the book issues pertinent to both these senses of 'importance' have been raised; now the threads will be drawn together.

6.1 AN ILLUSTRATION OF HOW MODERN SCIENCE WORKS

Less than ten years ago our knowledge of nucleocytoplasmic transport was aptly described as 'anecdotal' (Paine and Horowitz, 1980). Only in the last four years or so has the field been 'officially recognized', in the sense that symposia and conferences devoted to nucleocytoplasmic transport have been organized and major reviews have been published. Now a coherent, albeit provisional, picture is emerging. The field has an intrinsically limited conceptual scope (by definition of 'field'), but it has benefited from a wide range of experimental approaches. New techniques, and new ideas, have been imported to it from various other biological specialisms, and in some cases techniques have been invented within the field to address particular problems. For these reasons, it is very well suited to illustrate the processes by which modern science works.

Both *in situ* and *in vitro* methods have been used to analyse the structures relevant to nucleocytoplasmic transport, notably the NE, and the actual movements of macromolecules into and out of nuclei. Proponents of *in situ* methods and proponents of *in vitro* methods have shown an often contemptuous disregard of each others' work, at least so far as functional studies are concerned, very much as biochemists and cell biologists in general largely disregarded each other until around 1970. Despite this attitude, the two approaches complement each other, most obviously in the striking analogy between our present picture of cytoplasm-to-nucleoplasm protein transport, obtained largely by *in situ* methods, and that of nucleus-to-cytoplasm mRNA transport, obtained largely by *in vitro* methods. The 'Tower of Babel' phenomenon in which each small group of researchers pursues its own path and ignores others, unless it thinks it can instantly increase its publication rate by communicating with them, is all too common. It can make a laughing-stock of science and it needs to be resisted.

The nucleocytoplasmic transport field provides a vivid illustration of the fact that progress in science depends largely, but not wholly, on its methods. A method is useful just in so far as it enables us to answer a question that we want to answer, and our questions are expressed in terms of our current concepts, our mental image of the subject of study. Sometimes the availability of a method enables us to reformulate a question in a way that proves to be more helpful, and deployment of a method almost invariably leads to some surprising information: unexpected findings that enrich our conceptual repertoire and add detail to our mental pictures. This is the essence of progress in science: the movement towards a more and more coherent and detailed picture of the world, motivated by well-formulated questions, and achieved by the intimate interplay of concept and technique. This is the picture of the nucleocytoplasmic transport field presented in this book; it should be the picture of any field of research.

In the course of this book some very general points of discussion have been introduced: the meaning of the phrase 'well characterized', the use of analogies and counterfactuals in science, and so on. Also, individual scientific arguments have been reconstructed to emphasize the ways in which informed speculation can drive research and critical evaluation of methods, data and interpretations can guide it. Certainly, the vehicle of science can never progress far without the fuel of curiosity, the engine of speculative conjecture and the brakes and steering system of criticism. However, recognizing these points, involves the risk of over-rationalizing history. Scientists are prone to this fault; they are more or less trained to it. Even writing a paper involves a rational reconstruction of events that were often far from rationally ordered. In Section 5.1 the haphazard, often accidental-seeming way in which important insights *are* sometimes gained was contrasted with the process of rational thought by which they *could be* gained in principle.

Finally, throughout this book the aim has been to relate our nascent understanding of nucleocytoplasmic transport to our changing picture of the eukaryotic cell as a whole, and especially to the modern focus of interest in regulatory and developmental processes. For the rest of this chapter, these matters are discussed more directly.

6.2 NUCLEOCYTOPLASMIC TRANSPORT AND THE CONTROL OF CELL FUNCTION

6.2.1 Hormonal regulation

Many hormones act by switching the transcription of particular genes on or off. Most of the steroid hormones seem to act this way; which raises the question of how steroid hormones enter the target cell's nucleus. This process requires an intracellular protein, the steroid receptor.

Our opinions about the intracellular location of steroid receptors have undergone several shifts. A decade or so ago, everyone knew that the receptor proteins were normally cytoplasmic (because they were recovered in 100 000 g supernatants of tissue homogenates) and migrated rapidly into the nucleus when the steroid entered the cell and bound to them (because cell autoradiograms with tritium-labelled steroids showed an exclusively nuclear location). Then specific antibodies against the steroid receptors became available, which showed that they were always located in the nucleus, and at the same time we began to understand that massive leakage of nuclear proteins occurred during homogenization (Chapter 2). So opinion changed: now everyone knew that steroid receptors were exclusively nuclear proteins *in situ*, and the question arose of how the steroids found their way from the extracellular fluid through the cytoplasm to these receptors. Now opinions are changing again. The receptors' epitopes might be masked in the cytoplasm, so the antibodies do not reveal them in this compartment; and the kinetics of steroid uptake in cells with and without normal receptors certainly suggest the involvement of protein migration. We seem to be back to square one, with the important difference that now we have some idea of the mechanisms involved.

The glucocorticoid receptor in cultured monkey kidney cells seems to have two nuclear location signals, a situation that seems quite common among karyophilic proteins (Chapter 4). One of these signals, the more efficient of the two, is active (exposed?) only when the hormone is bound; the other on its own does not seem to be sufficient for nuclear uptake (Picard and Yamamoto, 1987). It therefore seems that hormone binding changes the tertiary structure of the receptor molecule so as to activate the crucial nuclear location signal. This is exactly the same mechanism as seems to be involved in U_2snRNP transport (Chapter 5). Perhaps the control of nuclear location by alterations of tertiary and quaternary structure is widespread in nature.

This example shows that nucleocytoplasmic transport processes can be involved in the intracellular interpretation of hormonal signals. It is easy to imagine mechanisms by which the signal itself can be modulated at this stage: for example, by modification of the receptor molecule so that the location signals remain masked even when the hormone binds. However, it is also possible that the machinery of nucleocytoplasmic transport itself can be controlled by hormones and other extracellular factors.

The possibility of controlling the mRNA translocation system via the protein kinase C in the NE (Chapter 5) is a case in point. These possibilities were reviewed recently (Agutter, 1988). Binding sites for macromolecules in the NS can also apparently be modified by the actions of hormones and second messengers (Sevaljevic *et al.*, 1982; Moy and Tew, 1986) and the possibilities this presents for alteration of nucleocytoplasmic distributions are obvious from the discussion in Section 4.1.

6.2.2 Carcinogen action

It is well established that carcinogens have mutagenic properties and act at the DNA level, e.g. by activation of proto-oncogenes. However, there may be supplementary mechanisms. The wealth of evidence that carcinogen treatment rapidly perturbes cytoplasmic mRNA populations without concomitant changes at the transcriptional level becomes very interesting given the rapid changes that occur in the mRNA translocation system in response to the same challenge (Shearer, 1974; Schumm *et al.*, 1977; Clawson *et al.*, 1980b). This could indicate a perturbation of cell metabolism at an early stage in carcinogenesis, perhaps a necessary prelude to the key changes at the DNA level. Also, the NE protein kinase C is sensitive to phorbol esters (Schröder *et al.*, 1988b). One possible contribution to the mechanism of action of co-carcinogens could therefore be alteration of the mRNA translocation machinery. Finally, carcinogens, like hormones, might alter intranuclear binding of transportable macromolecules rather than, or as well as, translocation processes. Rapidly-labelled RNA, for instance, seems to be more tightly bound to the NS in some cells than 'older' (more mature?) RNA (Zaboikin *et al.*, 1987; cf. also Schröder *et al.*, 1987). If this situation were altered by carcinogens, we could have an alternative explanation for the perturbation of cytoplasmic messenger populations by such reagents.

6.2.3 Regulatory processes in embryogenesis and ageing

During maturation of the oocyte, many nuclear proteins migrate to the cytoplasm at the time of germinal vesicle breakdown. Their distribution seems to depend on cytoplasmic movements, which of course are controlled by the cytoskeleton. After fertilization and during embroyogenesis, these proteins re-enter the nucleus in an apparently pre-programmed sequence (Hausen *et al.*, 1985). This process seems to be associated, presumably causally, with the programmed series of gene derepressions characteristic of early embryogenesis. Transcription does not begin until the mid-blastula transition, i.e. after 12 DNA replications and mitoses. The cytoplasm/ nucleus volume ratios in each of the daughter cells at this stage is much lower than the ratio in the egg. Newport and Kirschner (1982) suggested that a transcription suppressor or RNA polymerase suppressor is present in the cytoplasm; polymerase III, for example, is present from egg stage onwards but is inactive before the mid-blastula stage, so tRNA synthesis begins only at this point. As the relative volume of cytoplasm decreases, this hypothetical suppressor is effectively diluted and thus transcription begins.

Ageing is associated with a general running-down of metabolic processes, and once again nucleocytoplasmic transport processes might be involved causally. Certainly the mRNA translocation apparatus becomes less efficient: the NTPase activity decreases, the extent of phosphorylation of mRNA signal receptors in the NE declines and the poly(A) content of adenylated messengers is reduced (Müller *et al.*, 1985). Finally, sexual maturation in female quails is associated with a sudden increase in these transcription-related parameters in oviduct NE (Müller *et al.*, 1985). In general, nucleocytoplasmic transport processes seem to change concomitantly with various major stages in development, maturation and senescence.

6.2.4 Developmental changes in the nuclear envelope

Concomitantly with the changes in nucleocytoplasmic protein distributions and associated gene derepression in early embryogenesis in oocytes (see above), the biochemistry of the NE, at least of the lamina, undergoes changes. The lamins present in the cell are encoded on different genes at different stages of embryonic development (Krohne and Benavente, 1986). Work on this topic began from the observation that amphibian oocyte NEs contain only one (B-type) lamin, now known as L_{III} (Krohne *et al.*, 1978). More recent studies have identified a timed succession of lamins in the developing *Xenopus* NE; Krohne and his colleagues have cloned and partially sequenced the genes of several of these and they have identified the stages of development at which the various lamin genes are switched on and off (Krohne and Benavente, 1986). Studies of this kind are very important in developmental biology: they provide potentially valuable markers for clusters of genes that are presumably switched on by the same homeobox 'master genes' at appropriate developmental stages.

L_{III} disappears at or before the tadpole stage of development and appears in the adult only in a few specialized cells (neurons and muscle cells). The other gamete, the spermatozoon, also has a single lamin, L_{IV}, which is unique to this type of cell and appears late in spermatogenesis. Most somatic cells have two lamins, the A-type L_I with a molecular weight similar to that of L_{IV} (72–75 kDa) and the B-type L_{II} with the same molecular weight as L_{III} (68 kDa). There may be an analogous developmental pattern in mice, where a novel lamin which cross-reacts immunologically with the B-type lamin G of germinal vesicles of the clam *Spisula* appears uniquely during spermatogenesis (Maul *et al.*, 1986). This cross-reactivity between widely unrelated species indicates considerable evolutionary conservation. A timed succession of genes coding for closely homologous or virtually identical proteins at different stages of development is now a well-established pattern for several protein families; the lamins seem to fit into this pattern along with mammalian haemoglobins and sea-urchin histones.

The pertinence of these changes in the lamins to nucleocytoplasmic transport processes is not yet clear. It was suggested in Chapter 3 that the tight association between the lamina and the PCs might be relevant to the control of translocation processes. If this is so, then changes in the composition of the lamina might be prerequisites for changes in translocation capacity. The facts that lamins such as $L_{I/II}$ appear only when gene derepression begins, and the lamina consists exclusively of L_{III} or L_{IV} in cells where nucleocytoplasmic exchanges seem to be minimal, are certainly suggestive. In *Drosophila*, the lamina composition changes with metabolic conditions, notably in response to heat shock, which seems to impair nucleocytoplasmic transport processes, and of course many other cellular processes, fairly globally (Smith *et al.*, 1987). On the other hand, the importance of lamina composition might lie primarily in its capacity for chromatin binding and the control of transcription; and other as yet unidentified changes in the NE might coincide with changes in nucleocytoplasmic transport.

6.3 MECHANISMS

This discussion suggests that our improved understanding of nucleocytoplasmic transport will prove important in two of the main themes of modern cell biology: the mechanisms of control of cell function and of development. However, the potential relevance of the field goes beyond this. The picture of nucleocytoplasmic transport that we now have contains a few fairly novel features. These concepts might have wide implications for our image of the cell as a whole.

6.3.1 The solid-state model

At the beginning of the book it is stated that mRNA transport is at least a solid-state process, and since then this assumption has been tacitly accepted. Of course, the idea of a solid-state transport process is not new. The chromatids separate on the mitotic spindle, not in an unstructured medium, in most dividing cells; synaptic vesicles migrate along axons in apparently unbroken association with cytoskeletal fibres; and so on. What is new is the application of the model to a macromolecule or macromolecular complex rather than to an organelle. The evidence supporting the solid-state model has been discussed elsewhere (Agutter, 1985a, 1988); here some of the main points are summarized.

The tight association of HnRNP with the NS is now well-attested, and the snRNPs associated with it, e.g. in spliceosomes seem to be similarly tightly bound (Long *et al.*, 1979; Habets *et al.*, 1985; Smith *et al.*, 1986; Verheijen *et al.*, 1986). Association with the NS seems to begin at transcription itself (Jackson and Cook, 1985). The facts that splicing must precede translocation irrespective of messenger size; that distributions are not affected by NE puncturing; and that *in vitro* systems for studying mRNA efflux can exhibit near-normal nuclear restriction; all support the idea that messenger precursors are thoroughly non-diffusible in the nucleus. The existence of signal receptors in the NE suggests that the mRNA is not diffusible at the translocation stage of transport; and in the cytoplasm, all translationally active messengers are recovered in cytoskeletal preparations (van Vernrooij *et al.*, 1981), microinjected mRNA is rapidly immobilized (Drummond *et al.*, 1985) and mRNA does not redistribute during mitosis (Rao and Prescott, 1967). Clearly, mRNA and its precursors are always non-diffusible *in vivo*, though translationally inactive messengers in the cytoplasm might be exceptions to this rule. Messenger RNA efflux experiments don't work if the nuclei swell or if proteolysis or RNA degradation is allowed; any of these changes seems to disrupt the NS and thus 'solubilize' the HnRNP and intranuclear messengers from their attachment sites.

Three questions arise about the solid-state model: (1) How does solid-state transport work? (2) Do macromolecules other than mRNA move by a solid-state mechanism? (3) What advantage does a solid-state mechanism give to the cell?

1. The question 'how does it work?' is one facet of the problem of cytoskeletal dynamics. Gerd Maul (1982) suggested two mechanisms: a 'cable-car', in which the fibrils of the solid-state machinery move and carry the mRNA (or other) molecule attached to some part of them; and a 'rail-car', in which the fibrils are stationary and the mRNA (or whatever) travels along them with the aid of some kind of motor. Several recent

reviews of cytoskeletal mechanics in general are pertinent to these models in their general sense (Vale, 1987; Elson, 1988; Nicklas, 1988).

2. Protein transport is not usually considered a solid-state process: isolated nuclei don't seem to transport proteins in anything like a physiological way unless they have been resealed by Newmeyer's technique, and fluorescence micrographs of proteins microinjected into the cytoplasm show a diffuse distribution rather than a fibril-related one. The same is probably true of tRNA (Tobian *et al.*, 1984) and snRNPs (Mattaj and De Robertis, 1985); and snRNP migration into the nuclei does not seem to be affected by cytoskeleton-disrupting agents. Nevertheless, the work of Paine and others (Chapters 2 and 4) shows that most transportable proteins have at least restricted diffusibility, and the apparent distribution of the nuclear location signal receptors on cytoplasmic and perhaps intranuclear fibrils as well as the NE might mean that some solid-state processes are involved in nuclear permeation (Chapter 4). The general definition of 'transport' introduced in Chapter 1 might be relevant here: not just a matter of crossing the NE boundary, but movement between specific sites in the two major compartments. In the case of ribosomes, there is some evidence for tight association with the NS (Maul, 1977; Wunderlich *et al.*, 1984).

3. The teleological question is the most interesting one in the sense that a satisfactory answer would indicate a really satisfactory understanding. Of course, we do not have a satisfactory answer. Perhaps, by analogy with axonal transport, the solid-state mechanism ensures efficient vectorial transport. Perhaps the sustained association with skeletal structures in the cell helps to ensure the stability of messengers and to prevent their binding uselessly to other intracellular regions such as membrane surfaces. Perhaps it is necessary for efficient post-transcriptional processing and translation. Or perhaps it is some combination of these – or something else entirely.

Finally, there is as yet no good theoretical (biophysical) model of solid-state transport, and one is needed if these general questions are to be addressed satisfactorily. Some currently-evolving bodies of thought might generate such a model (Elson,1988; Nicklas, 1988; Kopelman, 1988). Once again, science is not just a matter of accumulating experimental data. Data are useless unless there is a body of theory capable of accommodating and organizing them.

6.3.2 How many signal receptors are there?

Our intuitions tell us that there must be a variety of receptors for the wide variety of signals involved in translocation processes. But is this really so? Colloidal gold particles coated with a karyophilic protein enter the nucleus by a PC through which particles coated with polynucleotides are leaving (Dworetzky and Feldherr, 1988), indicating that individual PCs are not specialized. The candidate receptors for protein signals that have been identified recently (Chapter 4) might include the many reproducible fragments of the P110 mRNA signal receptor; at least they have the same molecular weights and something like the same intracellular distribution. Also, recent evidence from Fasold's laboratory (personal communication) shows that histones entering resealed NE vesicles cross-link to a series of polypeptides with the same molecular weights as the P110 degradation products. Is it possible that there is just one multifunctional signal receptor molecule involved in translocation processes? The idea is certainly counter-intuitive, but to take an analogy: actin exhibits specific, saturable, high-affinity binding to a very large range of other proteins, and actin is a much smaller protein than P110 (whatever the real molecular weight of this object is). It is better not to press this point for fear of using analogy as justification, not heuristic. Good antibody evidence will be needed even to make the matter worth investigating.

Nucleocytoplasmic exchanges of proteins and of mRNAs are blocked by different antibodies (Sugawa and Uchida, 1985). This could support the concept of a family of receptors rather than a single multifunctional one, but on the other hand the antibodies might interfere with intracompartmental binding rather than translocation, and there is no need to suggest that there is a single intranuclear binding site for all karyophilic proteins and nuclear RNAs!

6.3.3 Why are the affinities of the signal receptors so low?

In all cases in which the affinity of the receptor for the signal has been measured (proteins, mRNA and tRNA) the dissociation constant is in the region of 10^{-7} M. What sense can we make of this information?

In the case of mRNA, the low affinity of the receptor could make the translocation machinery act as an amplifier of abundance differences. Suppose we have two messengers, X and Y, in the nucleus. Suppose X has a concentration less than 10^{-7} M and Y has a concentration above this magic value. (The concept of 'concentration' in a solid-state model needs some explication, but the term can be used, albeit metaphorically, without confusing the issue.) The receptor's low affinity will favour the binding of Y rather than X. This means that the probability that X will be moved to the cytoplasm in any translocation event is much lower than the probability of translocating Y. Perhaps this explains why the differences between messenger abundance classes are so much greater in the cytoplasm than in the nucleus: it is (the solid-state equivalent of) a simple mass-action effect. Moreover, if the receptor's affinity is increased by protein kinase C dependent phosphorylation, for example in response to phorbol ester exposure, this amplifying effect will be impaired. This could be pertinent to the 'scrambling' of abundance class distinctions that happens in carcinogenesis (page 121). These arguments don't explain the importance for the cell of the receptor's low affinity, but at least they relate it to well-known biological phenomena.

What about proteins? In the case of high-abundance proteins such as nucleoplasmin, the affinity for the receptor scarcely matters; the receptor will be saturated anyway. For low-abundance proteins the problem appears more significant. Norbert Riedel (personal communication) suggests that the low affinity indicates a rapid off-rate from the ligand–receptor complex, making dissociation more efficient; but this is unlikely to facilitate translocation because the rate limiting step in nuclear permeation is not the interaction between ligand and receptor, but the translocation event *per se* (Chapter 4). On the other hand, if there are very few receptor molecules, say a few per PC, therefore in the order of 10^4 per rat liver nucleus, binding of a low-abundance ligand will result in few ligand–receptor complexes and therefore translocation must be very inefficient. Translocation is certainly slow. Generally accepted figures for the maximum rate of export of ribosomal subunits are around 5–6 per PC per minute; and even the most rapid translocation processes, such as histone uptake, seem not to go faster than 100 molecules per PC per minute, or 1–2 molecules per PC per second.

This brings us back to the significance of the intracellular distribution of signal receptors. If they really are present on an extensive fibrillar network as well as in the NE, then there are many more receptor molecules per cell than there would be if they were confined to the NE alone. At any instant, this would greatly increase the number of ligand–receptor complexes for a low-abundance ligand, despite the high dissociation constant value. If the solid-state machinery of the cell were geared to deliver these complexes to the PC and translocation was then in effect vectorial (e.g. because of intranuclear binding), then the approximately 'all-or-nothing' effect on nuclear permeation observed for wild-type and defective signals might be predicted (Chapter 4). This 'all-or-nothing' effect cannot be achieved by, for example resealed NE vesicles; perhaps this is because the solid-state machinery is missing. This hypothesis is not necessarily inconsistent with the cytoplasmic fluorescence pictures of microinjected cells – free ligand will predominate at virtually any time until nuclear uptake is complete – and the idea is testable: nuclear protein uptake should be susceptible to significant retardation by agents that disrupt the cytoskeleton.

6.4 EVOLUTIONARY CONSIDERATIONS

Nucleocytoplasmic transport is perhaps the fundamental distinctively eukaryotic process. A eukaryotic cell, by definition, is divided into nucleus and cytoplasm throughout interphase. The evolution of the first eukaryote must therefore have involved not only the 'invention' of the NE, but also of a whole apparatus for specific, controlled nucleocytoplasmic exchange of macromolecules. No doubt the components of this apparatus were recruited from our nearest prokaryotic ancestors and the whole has been fine-tuned by subsequent evolutionary developments, but it is quite clear that PCs, intramolecular signals, signal receptors and a solid-state machinery must have been present in the earliest eukaryotic cell. How did they come into being, and what advantages did they confer on the cell?

Tom Cavalier-Smith points out that the evolution of eukaryotes from prokaryotes might be as serious a challenge to our understanding as the origin of life, and he has devoted much of the last 15 years to constructing a theory of eukaryote evolution. Comparing some of his many publications in this field over this period provides a good illustration of the way in which scientific ideas can be progressively refined (e.g. Cavalier-Smith, 1975, 1981, 1987, 1988). Granted that 'origin of eukaryotes' = 'origin of separate nucleus', he identifies six specific problems for his theory to address: (a) origin of ER, ONM and INM from the prokaryotic plasma membrane; (b) modifications of chromosome structure, particularly the appearance of centromeres, telomeres, nucleosomes and multiple replication sites; (c) origin of mitosis; (d) origin of meiosis and the synaptonemal complex; (e) origin of post-transcriptional RNA processing and spliceosomes; (f) origin of the PC, NS, and location or translocation signals (Cavalier-Smith, 1988). These six problems are obviously interrelated. Briefly, Cavalier-Smith argues that DNA itself is the primary structural organizer of the nucleus (cf. Forbes *et al.*, 1983), and the first eukaryote evolved from a wall-less prokaryote with a primitive cytoskeleton and DNA attachment sites on the plasma membrane that could function as centromeres. The primitive cytoskeleton included lamin/cytokeratin-like proteins which interfaced the DNA with the internalized membrane, thus forming the first NE, and actin and myosin to provide a primitive motor for segregation of the internalized DNA in replication events. The PC must have developed, according to this hypothesis, from proteins associated with this primitive cytoskeleton (cf. the modern-day tight binding of PCs to lamin B: Chapter 3). Cavalier-Smith also suggests that the NTPase was primitively associated with the PC and has evolved from a prokaryotic membrane-bound mechanochemical ATPase, the evolutionary ancestor of myosin, kinesin, dynein and other cytoskeletal motors.

As for the advantages of the eukaryotic state, Cavalier-Smith lists (a) protection of the DNA from mechanical damage caused by cytoplasmic contractility, (b) ensuring more or less equal segregation of functional parts to the daughter cells by attachment (albeit indirect) of the DNA to the endomembranes, and (c) concentration of the macromolecular assemblies necessary for different metabolic purposes in more confined subcellular spaces, thus increasing efficiency. Note that point (c) is valid only if we assume the inevitability of a substantial increase of total cell volume over that found in prokaryotes. It is an interesting exercise to try to add to this list.

The ability to elaborate this kind of model is a good test of our understanding, and is recommended to the reader as an armchair puzzle or discussion topic. One thing needs to be kept in mind: the remarkable evolutionary conservation of the fundamental apparatus (PC, lamins, etc.). Since the machinery was formed it seems to have undergone no fundamental change, in contrast to the machineries of other major nuclear phenomena such as mitosis, which involves complete NE breakdown only in organisms higher than fungi. Antibodies against lamin A, lamin B and the lamin B receptor in turkeys cross-react with yeast NE (Georgatos *et al.*, 1989), and similar antibody evidence confirms the remarkable conservation of PC polypeptide components between yeasts and mammals (Aris and Blobel, 1989). Evidence of this kind emphasizes how fundamental to the eukaryotic state are the structures of the NE and the mechanisms of exchange that we observe today between nucleus and cytoplasm.

References

Aaronson, R.P. and Blobel, G. (1975) Isolation of nuclear pore complexes in association with a lamina. *Proc. Natl Acad. Sci. USA*, **72**, 1007–11.

Abelson, H.T. and Smith, G.H. (1970) Nuclear pores: the pore-annulus relationship in thin sectioning. *J. Ultrastr. Res.*, **30**, 558–88.

Adam, S.A., Lobl, T.J. Mitchell, M.A. *et al.* (1989) Identification of specific binding proteins for a nuclear location sequence. *Nature*, **337**, 176–9.

Aebi, U., Kohn, J., Buhle, L. *et al.* (1986) The nuclear lamina is a meshwork of intermediate-type filaments. *Nature*, **323**, 560–4.

Agutter, P.S. (1985a) RNA processing, RNA transport and nuclear structure, in *The Nuclear Envelope and RNA Maturation*, UCLA Symposium, Vol. 26 (eds G.A. Clawson and E.A. Smuckler), Alan R. Liss, New York, pp. 539–60.

Agutter, P.S. (1985b) The nuclear envelope NTPase and RNA translocation, in *The Nuclear Envelope and RNA Maturation*, UCLA Symposium, Vol. 26 (eds G.A. Clawson and E.A. Smuckler), Alan R. Liss, New York, pp. 561–78.

Agutter, P.S. (1988) Nucleo-cytoplasmic transport of mRNA: its relationship to RNA metabolism, subcellular structures and other nucleocytoplasmic exchanges, in *Prog. Mol. Subcell. Biochem.*, Vol. 10 (ed. W.E.G. Müller), Springer-Verlag, Heidelberg, pp. 15–96.

Agutter, P.S. (1990) Export of poly(A)+RNA from resealed rat liver nuclear envelope vesicles. *Biochem. J.*, in press.

Agutter, P.S., Harris, J.R. and Stevenson, I. (1977) RNA stimulation of mammalian liver nuclear envelope NTPase. *Biochem. J.*, **162**, 671–9.

Agutter, P.S., McArdle, H.J. and McCaldin, B. (1976) Evidence of involvement of nuclear envelope NTPase in nucleocytoplasmic translocation of ribonucleoproteins. *Nature*, **263**, 165–7.

Agutter, P.S. and Suckling, K.E. (1982) The fluidity of the nuclear envelope lipid does not affect the rate of nucleocytoplasmic RNA transport in mammalian liver. *Biochem. Biophys. Acta*, **696**, 308–14.

Archega, J. and Bahr, G.F. (1985) in *The Nuclear Envelope and RNA Maturation*, UCLA Symposium Vol. 26 (eds G.A. Clawson and E.A. Smuckler) Alan R. Liss, New York, pp. 23–50.

Arenstorf, H.P., Conway, G.C., Wooley, J.C. *et al.* (1984) Nuclear matrix-like filaments form through artefactual rearrangements of HnRNP particles. *J. Cell Biol.*, **99**, 233a.

Aris, J.P. and Blobel, G. (1989) Yeast nuclear envelope proteins cross-react with an antibody against mammalian pore complex proteins. *J. Cell Biol.*, **108**, 2059–68.

Austerberry, C.F. and Paine, P.L. (1982) *In vivo* distribution of proteins within a single cell. *Clin. Chem.*, **28**, 1011–14.

Baglia, F.A. and Maul, G.G. (1983) Nuclear ribonucleoprotein release and NTPase activity are inhibited by antibodies directed against one nuclear matrix glycoprotein. *Proc. Natl Acad. Sci. USA*, **80**, 2285–98.

Beck, J.S. (1962) The behaviour of certain nuclear antigens in mitosis. *Exp. Cell Res.*, **28**, 406–18.

Benavente, R. and Krohne, G. (1986) Involvement of nuclear lamins in post-mitotic reorganization of chromatin as demonstrated by microinjection of lamin antibodies. *J. Cell Biol.*, **103**, 1847–54.

Benditt, J.O., Meyer, C., Fasold, H. *et al.* (1990) Interaction of a nuclear location signal with isolated nuclear envelopes and identification of signal binding proteins by photoaffinity labeling. *Proc. Natl Acad. Sci. USA*, in press.

Berezney, R. (1979) Dynamics of the nuclear protein matrix, in *The Cell Nucleus*, Vol. 7 (ed. H. Busch), Academic Press, New York and London, pp. 413–55.

Berezney, R. and Coffey, D.S. (1974) Identification of a nuclear protein matrix. *Biochem. Biophys. Res. Commun.*, **60**, 1410–17.

Bernd, A., Schröder, H.-C., Zahn, R.K. *et al.* (1982) Modulation of the nuclear envelope NTPase by poly(A) rich mRNA and by microtubule proteins. *Eur. J. Biochem.*, **129**, 43–9.

Berrios, M. and Fisher, P.A. (1986) A myosin heavy chain-like polypeptide is associated with the nuclear envelope in higher eukaryotic cells. *J. Cell Biol.*, **103**, 711–24.

Berrios, M., Osherhoff, N. and Fisher, P.A. (1985) *In situ* localization of DNA topoisomerase II, a major polypeptide component of the *Drosophila* nuclear matrix. *Proc. Natl Acad. Sci. USA*, **82**, 4142–46.

Birnie, G.D. and MacGillivray, A.S. (eds) (1986) *Nuclear Structures: their Isolation and Characterisation*, Butterworths, London.

Bladon, T., Brasch, K.R., Brown, D.L. *et al.* (1988) Changes in structure and protein composition of bovine lymphocyte nuclear matrix during concanavalin-A-stimulated mitogenesis. *Biochem. Cell Biol.*, **66**, 40–53.

Blobel G. (1973) A protein of molecular weight 73 000 bound to the polyadenylate regions of eukaryotic messenger RNAs. *Proc. Natl Acad. Sci. USA*, **70**, 924–8.

Blobel, G. and Potter, V.R. (1966) Nuclei from rat liver: isolation method that combines purity with high yield. *Science*, **154**, 1662–5.

Bonner, W.M. (1975a) Protein migration into nuclei I. Frog oocyte nuclei *in vivo* accumulate microinjected histones, allow entry to small proteins, and exclude large proteins. *J. Cell Biol.*, **64**, 421–30.

Bonner, W.M. (1975b) Protein migration into nuclei II. Frog oocyte nuclei accumulate a class of microinjected oocyte nuclear proteins and exclude a class of microinjected cytoplasmic proteins. *J. Cell Biol.*, **64**, 431–7.

Bornens, M. and Courvalin, J.C. (1978) Isolation of nuclear envelopes with polyanions. *J. Cell Biol.*, **76**, 191–206.

Brasch, K. (1982) Fine structure and localisation of the nuclear matrix *in situ*. *Exp. Cell Res.*, **140**, 161–72.

Bürglin, T.R. and de Robertis, E.M. (1987) The nuclear migration signal of *Xenopus laevis* nucleoplasmin. *EMBO J.*, **6**, 2617–25.

Burgoyne, L.A. and Skinner, J.D. (1979) Probing the free space within rat and chicken chromatin with active and passive probes. *J. Cell Sci.*, **37**, 85–96.

Burke, B. and Gerace, L. (1986) A cell free system to study reassembly of the nuclear envelope at the end of mitosis. *Cell*, **44**, 639–52.

Callan, H.G., Randall, J.T. and Tomlin, S.G. (1949) An electron microscope study of the nuclear membrane. *Nature (Lond.)*, **163**, 280.

Capco, D.G. and Penman, S. (1983) Mitotic architecture of the cell: the filament networks of the nucleus and cytoplasm. *J. Cell Biol.*, **96**, 896–906.

Carmo-Fonseca, M., Cicadão, A.J. and David-Ferreira, J.F. (1987) Filamentous cross-bridges link intermediate filaments to the nuclear pore complexes. *Eur. J. Cell Biol.*, **45**, 282–90.

Carter, C.J., Cannon, M. and Jiménez, A. (1980) A trichodermin-resistant mutant of *Saccharomyces cerevisiae* with an abnormal distribution of ribosomal subunits. *Eur. J. Biochem.*, **107**, 173–83.

Cavalier-Smith, T. (1975) The origin of nuclei and of eukaryotic cells. *Nature (Lond.)*, **265**, 463–8.

Cavalier-Smith, T. (1983) The origin and early evolution of the eukaryotic cell. *Symp. Soc. Gen. Microbiol.*, **32**, 33–84.

Cavalier-Smith, T. (1987) The origin of eukaryote and archebacterial cells. *Ann. NY Acad. Sci.*, **503**, 17–54.

Cavalier-Smith, T. (1988) Origin of the cell nucleus. *BioEssays*, **9**, 72–8.

Chaly, N., Bladon, T., Setterfield, G. *et al.* (1984) Changes in distribution of nuclear matrix antigens during the mitotic cell cycle. *J. Cell Biol.*, **99**, 661–71.

Chan, P-K., Chan, W-Y., Yung, B.Y.M. *et al.* (1986) Amino acid sequence of a specific antigenic peptide of protein B23. *J. Biol. Chem.*, **261**, 14335–41.

Clawson, G.A., James, J., Woo, C.H. *et al.* (1980a) Pertinence of nuclear envelope NTPase activity to transport of RNA from rat liver nuclei. *Biochemistry*, **19**, 2756–62.

Clawson, G.A., Koplitz, M., Moody, D.E. *et al.* (1980b) Effects of thioacetamide treatment on nuclear envelope NTPase activity and transport of RNA from rat liver nuclei. *Cancer Res.*, **40**, 75–9.

Comerford, S.A., Agutter, P.S. and McLennan, A.G. (1986) Nuclear matrices, in *Nuclear Structures: their Isolation and Characterization* (eds A.J. MacGillivray and G.D. Birnie), Butterworths, London, pp. 1–13.

Dabauville, M.C. and Franke, W.W. (1982) Karyophilic proteins: polypeptides synthesized *in vitro* accumulate in the nucleus upon microinjection into the cytoplasm of amphibian oocytes. *Proc. Natl Acad. Sci. USA*, **79**, 5302–6.

Davis, L.I. and Blobel, G. (1986) Identification and characterization of a nuclear pore complex protein. *Cell*, **45**, 699–709.

De la Peña, P. and Zasloff M. (1987) Enhancement of mRNA nuclear transport by promoter elements. *Cell*, **50**, 613–19.

De Robertis, E.M. (1983) Intracellular migration of nuclear proteins in *Xenopus* oocytes. *Cell*, **32**, 1021–5.

De Robertis, E.M., Longthorne, R.F. and Gurdon, J.B. (1978) Intracellular migration of nuclear proteins in *Xenopus* oocytes. *Nature (Lond.)*, **272**, 254–6.

Dingwall, C. and Laskey, R.A. (1986) Protein import into the cell nucleus. *Annu. Rev. Cell Biol.*, **2**, 367–90.

Dingwall, C., Robbins, J., Dilworth, S.M. *et al.* (1988) The nucleoplasmin nuclear location signal is larger and more complex than that of SV_{40} large-T antigen. *J. Cell Biol.*, **107**, 841–850.

Dingwall, C., Sharnick, S.V. and Laskey, R.A. (1982) A polypeptide domain that specifies migration of nucleoplasmin into the nucleus. *Cell*, **30**, 449–58.

Dreyer, C., Stick, R. and Hausen, P. (1986) Uptake of oocyte proteins by nuclei of *Xenopus* embryos, in *Nucleo-cytoplasmic Transport* (eds R. Peters and M. Trendelenberg), Springer-Verlag, Heidelberg, pp. 143–57.

Dreyer, C., Wang, Y-H., Wedlich, D. *et al.* (1983) Oocyte nuclear proteins in the development of *Xenopus*, in *British Society for Developmental Biology Symposium* (eds P. Hausen and A. McLaren), Cambridge University Press, London, pp. 285–331.

Drummond, D.R., McCrae, M.A. and Colman, A. (1985) Stability and movement of mRNAs and their encoded proteins in *Xenopus* oocytes. *J. Cell Biol.*, **100**, 1148–56.

Dworetzky, S.I. and Feldherr, C.M. (1988) Translocation of RNA-coated gold particles through the nuclear pores of hepatocytes. *J. Cell Biol.*, **106**, 575–84.

Dworetzky, S.I., Lanford, R.E. and Feldherr, C.M. (1988) The effects of variations in the number and sequence of targeting signals on nuclear utake. *J. Cell Biol.*, **107**, 1279–88.

Elson, E.L. (1988) Cellular mechanics as an indicator of cytoskeletal structure and function. *Ann. Rev. Biophys. Biophys. Chem.*, **17**, 397–430.

Fawcett, D.W. (1966) On the occurrence of a fibrous lamina on the inner aspect of the nuclear envelope in certain cells of vertebrates. *Am. J. Anat.*, **119**, 129–46.

Featherstone, C., Darby, M.K. and Gerace, L. (1988) A monoclonal antibody against the nuclear pore complex inhibits nucleocytoplasmic transport of protein and RNA *in vivo*. *J. Cell Biol.*, **107**, 1289–98.

Feldherr, C.M. (1962) The nuclear annuli as pathways for nucleocytoplasmic exchanges. *J. Cell Biol.*, **14**, 65–72.

Feldherr, C.M. (1972) Structure and function of the nuclear envelope, in *Advances in Cell and Molecular Biology* (ed. E.J. Dupraw), Academic Press, London and New York, pp. 273–307.

Feldherr, C.M., Kallenbach, E. and Schultz, N. (1984) Movement of a karyophilic protein through the nuclear pores of oocytes. *J. Cell Biol.*, **99**, 2216–22.

Feldherr, C.M. and Pomerantz, J. (1978) Mechanism for the selection of nuclear polypeptides in *Xenopus* oocytes. *J. Cell Biol.*, **78**, 168–75.

Fenichel, I.R. and Horowitz, S.B. (1969) Intracellular transport, in *Biological Membranes* (ed. R.M. Dowben), Little Brown, Boston, pp. 177–221.

Fey, E.G., Krochmalnic, G. and Penman, S. (1986) The nonchromatin substructures of the nucleus: the RNP-containing and RNP-depleted matrices analysed by sequential fractionation and resinless electron microscopy. *J. Cell Biol.*, **102**, 1654–65.

Fey, E.G., Wan, K.M. and Penman, S. (1984) Epithelial cytoskeletal framework and nuclear matrix/intermediate filament scaffold: three-dimensional organization and protein composition. *J. Cell Biol.*, **98**, 1973–84.

Fields, A.P., Kaufmann, S.H. and Shaper, J.H. (1986) Analysis of the internal nuclear matrix: oligomers of a 38 kD nucleolar polypeptide stabilized by disulfide bonds. *Exp. Cell Res.*, **164**, 139–53.

Finlay, D.R., Newmeyer, D.D., Price, T.M. and Forbes, D.J. (1987) Inhibition of *in vitro* nuclear transport by a lectin that binds to nuclear pores. *J. Cell Biol.*, **104**, 189–200.

Forbes, D.J., Kirschner, M.W. and Newport, J.W. (1983) Spontaneous formation of nucleus-like structures around bacteriophage DNA microinjected into *Xenopus* eggs. *Cell*, **34**, 13–23.

Franke, W.W. (1974) Structure, biochemistry and functions of the nuclear envelope. *Int. Rev. Cytol.* Suppl. **4**, 72–236.

Franke, W.W. (1987) Nuclear lamins and cytoplasmic intermediate filament proteins: a growing multigene family. *Cell*, **18**, 33–4.

Franke, W.W., Scheer, U., Krohne, G. and Jarasch, E.-J. (1981) The nuclear envelope. *J. Cell Biol.*, **91**, 39s–50s.

Fulton, A.B. (1983) How crowded is the cytoplasm? *Cell*, **30**, 345–7.

Gall, J.G. (1964) Pore complexes. *Protoplasmatologia*, **5**, 4–34.

Georgatos, S.D. and Blobel, G. (1987) Lamin B constitutes an intermediate filament attachment site at the nuclear envelope. *J. Cell Biol.*, **105**, 117–25.

Georgatos, S.D., Maroulakou, I. and Blobel, G. (1989) Lamin A, lamin B and lamin B receptor analogues in yeast. *J. Cell Biol.*, **108**, 2069–82.

Georgatos, S.D., Stournaras, C. and Blobel, G. (1988) Heterotypic and homotypic associations between the nuclear lamins: site-specificity and control by phosphorylation. *Proc. Natl Acad. Sci. USA*, **85**, 4325–9.

Gerace, L. and Blobel, G. (1980) The nuclear envelope lamina is reversibly depolymerized during mitosis. *Cell*, **19**, 277–88.

Gerace, L., Ottaviano, Y. and Kondor-Koch, C. (1982) Identification of a major polypeptide of the nuclear pore complex. *J. Cell Biol.*, **95**, 826–37.

Goldfarb, D.S., Gariepy, J., Schoolnik, G. *et al.* (1986) Synthetic peptides as nuclear localization signals. *Nature (Lond.)*, **322**, 641–4.

Goldfine, I.D., Clawson, G.A., Smuckler, E.A. *et al.* (1982) Action of insulin at the nuclear envelope. *Mol. Cell Biochem.*, **48**, 3–14.

Gruss, P., Lai, C.-J., Dhar, R. *et al.* (1979) Splicing as a requirement for biogenesis of functional 16S mRNA of SV$_{40}$. *Proc. Natl Acad. Sci. USA*, **76**, 4317–21.

Gurdon, J.B. (1970) Nuclear transplantation and the control of gene activity during animal development. *Proc. R. Soc. Lond. (Biol.)*, **176**, 303–14.

Gurr, M.I., Finean, J.B. and Hawthorne, J.N. (1963) The phospholipids of liver-cell fractions. I: the phospholipid composition of the liver-cell nucleus. *Biochim. Biophys. Acta*, **70**, 406–16.

Habets, W.J., Berden, J.H.M., Hoch, S.O. *et al.* (1985) Further characterization and subcellular localization of Sm and U1 ribonucleoprotein antigens. *Eur. J. Immunol.*, **15**, 992–7.

Hall, M.N., Hereford, L. and Herskowitz, I. (1984) Targeting of *Escherichia coli* β-galactosidase to the nucleus in yeast. *Cell*, **36**, 1057–65.

Hausen, P., Wang, Y-H., Dreyer, C. *et al.* (1985) Distribution of nuclear proteins during maturation of the *Xenopus* oocyte. *J. Embryol. Exp. Morphol.*, **89**, Suppl. 17–34.

Holt, G.D., Snow, C.M., Senior, A. *et al.* (1987) Nuclear pore complex glycoproteins contain cytoplasmically disposed O-linked N-acetylglucosamine. *J. Cell Biol.*, **104**, 1157–64.

Horowitz, S.B. and Paine, P.L. (1976) Cytoplasmic exclusion as a basis for asymmetric nucleocytoplasmic solute distributions. *Nature (Lond.)*, **260**, 151–3.

Jackson, D.A. and Cook, P.R. (1985) Transcription occurs at a nucleoskeleton. *EMBO J.*, **4**, 919–26.

Jelinek, W. and Goldstein, L. (1973) Isolation and characterization of some of the proteins that shuttle between nucleus and cytoplasm in *Ameba proteus*. *J. Cell Physiol.*, **81**, 181–97.

Jones, N.L. and Kirkpatrick, B.A. (1988) The effects of human cytomegalovirus infection on cytoskeleton-associated polysomes. *Eur. J. Cell Biol.*, **46**, 31–8.

Kalderon, D., Richardson, W.D., Markham, A.F. *et al.* (1984) Sequence requirements for nuclear location of SV$_{40}$ large-T antigen. *Nature (Lond.)*, **311**, 33–8.

Kaufman, S.H., Coffey, D.S. and Shaper, J.H. (1981) Considerations in the isolation of rat liver nuclear matrix, nuclear envelope and pore complex lamina. *Exp. Cell Res.*, **132**, 105–23.

Kaufman, S.H., Gibson, W. and Shaper, J.H. (1983) Characterization of the major polypeptides of the rat liver nuclear envelope. *J. Biol. Chem.*, **258**, 2710–19.

Kay, R.R., Fraser, D. and Johnston, I.R. (1972) A method for the rapid isolation of nuclear membranes from rat liver. *Eur. J. Biochem.*, **30**, 145–54.

Kirschner, R.M., Rusli, M. and Martin, T.E. (1977) Characterization of the nuclear envelope, pore complex and dense lamina of mouse liver nuclei at high resolution by scanning electron microscopy. *J. Cell Biol.*, **72**, 118–32.

Kleinschmidt, J.A. and Seiter, A. (1988) Identification of domains involved in nuclear uptake and histone binding of protein N1 of *Xenopus laevis*. *EMBO J.*, **7**, 1605–14.

Kondor-Koch, C., Riedel, N., Valentin, R. *et al.* (1982) Characterization of an ATPase on the inside of rat liver nuclear envelopes by affinity labelling. *Eur. J. Biochem.*, **127**, 285–90.

Konings, D.A.M. and Mattaj, I.W. (1987) Mutant U_2-snRNAs of *Xenopus* which can form an altered higher order RNA structure are unable to enter the nucleus. *Exp. Cell Res.*, **172**, 329–39.

Kopelman, R. (1988) Fractal reaction kinetics. *Science*, **241**, 1620–6.

Krachmarov, C., Tasheva, B., Markov, D. *et al.* (1986) Isolation and characterization of nuclear lamina from Ehrlich ascites tumor cells. *J. Cell Biochem.*, **30**, 351–9.

Krohne, G. and Benavente, R. (1986) The nuclear lamins: a multigene family of proteins in evolution and differentiation. *Exp. Cell Res.*, **162**, 1–10.

Krohne, G., Franke, W.W., Ely, A. *et al.* (1978) Localization of a nuclear-envelope associated protein by indirect immunofluorescence microscopy using antibodies against a major polypeptide from rat liver fractions enriched in nuclear envelope-associated material. *Cytobiologie*, **18**, 22–38.

Krohne, G., Stick, R., Kleinschmidt, J.A. *et al.* (1972) Immunological localization of a major karyoskeletal protein in nucleoli of oocytes and somatic cells of *Xenopus laevis*. *J. Cell Biol.*, **94**, 749–54.

Laliberté, J-F, Dagenais, A., Fillion, M. *et al.* (1984) Identification of distinct messenger RNAs for nuclear lamin C and a putative precursor of nuclear lamin A. *J. Cell Biol.*, **98**, 980–5.

Lanford, R.E., Kanda, P. and Kennedy, R.C. (1986) Induction of nuclear transport with a synthetic peptide homologous to the SV40 T antigen transport signal. *Cell*, **46**, 575–82.

Lanford, R.E., White, R.G., Dunham, R.G. *et al.* (1988) Effect of basic and nonbasic amino-acid substitutions on transport induced by SV_{40} large-T antigen synthetic peptide nuclear transport signals. *Mol. Cell Biol.*, **8**, 2722–9.

Lang, I. and Peters, R. (1984) Nuclear envelope permeability: a sensitive indicator of pore-complex integrity. *Prog. Clin. Biol. Res.*, **164**, 377–86.

LeStourgeon, W.M. (1978) The occurrence of contractile proteins in nuclei and their possible functions, in *The Cell Nucleus – Chromatin, Part C*, (Vol. VI) (ed. H. Busch), Academic Press, New York, pp. 305–26.

Lohka, M.J. and Masui, Y., (1983) Formation *in vitro* of sperm pronuclei and mitotic chromosomes by amphibian ooplasmic components. *Science*, **220**, 719–21.

Long, B.H., Wang, C-Y. and Pogo, A.O. (1979) Isolation and characterization of the nuclear matrix in Friend erythroleukemia cells: chromatin and HnRNA interactions with the nuclear matrix. *Cell*, **18**, 1079–90.

McDonald, J.R. and Agutter, P.S. (1980) The relationship between polyribonucleotide binding and the phosphorylation and dephosphorylation of nuclear envelope protein. *FEBS Lett.*, **116**, 145–8.

McKeon, F.D., Kirschner, M.W. and Caput, D. (1986) Homologies in both primary and secondary structure between nuclear envelope and intermediate filament proteins. *Nature*, **319**, 463–8.

Madsen, P., Nielsen, S. and Celis, J.E. (1983) Monoclonal antibody specific for human nuclear proteins IEF 8Z30 and 8Z31 accumulates in nuclei a few hours after cytoplasmic microinjection of cells expressing these proteins. *J. Cell Biol.*, **103**, 2083–9.

Marsden, M.P.F. and Laemmli, U.K. (1979) Metaphase chromosome structure: evidence for a radial loop model. *Cell*, **16**, 849–58.

Mattaj, I.W. and De Robertis, E.M. (1985) Nuclear segregation of U2snRNA requires binding of specific snRNP proteins. *Cell*, **40**, 111–18.

Maul, G.G. (1977) The nuclear and cytoplasmic pore complex: structure, dynamics, distribution and evolution. *Int., Rev. Cytol.*, Suppl. **6**, 75–186.

Maul, G.G. (1982) *The Nuclear Envelope and the Nuclear Matrix*, Wistar Symposium Series Vol. 2, Alan R. Liss, New York.

Maul, G.G. and Baglia, F.A. (1983) Localization of a major nuclear envelope protein by differential solubilization. *Exp. Cell Res.*, **145**, 285–92.

Maul, G.G., French, B.T. and Bechtol, K.B. (1986) Identification and redistribution of lamins during nuclear differentiation in mouse spermatogenesis. *Dev. Biol.*, **115**, 68–77.

Moffett, R.B. and Webb, T.E. (1983) Characterization of a messenger RNA transport protein. *Biochim. Biophys. Acta*, **740**, 231–42.

Moy, B.C. and Tew, K.D. (1986) Differences in the nuclear matrix phosphoproteins of a wild-type and nitrogen mustard-resistant rat breast carcinoma cell line. *Cancer Res.*, **46**, 4672–6.

Müller, W.E.G., Agutter, P.S., Bernd, A. *et al.* (1985) Role of post-transcriptional events in aging: consequences for gene expression in eukaryotic cells, in *Thresholds in Aging* (eds M. Bergener, M. Erminici and H.B. Stählin), 1984 Sandoz Lectures in Gerontology, Academic Press, London, pp. 21–58.

Murty, C.N., Verney, E. and Sidransky, H. (1980) Effect of tryptophan on nuclear envelope nucleoside triphosphatase in rat liver. *Proc. Soc. Exp. Biol. Med.*, **163**, 155–61.

Nakayasu, H. and Ueda, K. (1984) Small nuclear RNP complex anchors on the actin filaments in bovine lymphocyte nuclear matrix. *Cell Struct. Funct.* **9**, 317–26.

Newmayer, D.D. and Forbes, D.J. (1988) Nuclear import can be separated into two distinct steps *in vivo*: nuclear pore binding and translocation. *Cell*, **52**, 641–53.

Newmayer, D.D., Lucocq, J.M., Bürglin, T.R. *et al.* (1986) Assembly *in vitro* of nuclei active in nuclear protein transport: ATP is required for nucleoplasmin accumulation. *EMBO J.*, **5**, 501–10.

Newport, J. (1987) Nuclear reassembly *in vitro*: stages of assembly around protein-free DNA. *Cell*, **48**, 205–17.

Newport, J. and Kirschner, M.W. (1982) A major developmental transition in early *Xenopus* embryos. *Cell*, **30**, 675–96.

Newport, J. and Spann, T. (1987) Disassembly of the nucleus in mitotic extracts: membrane vesicularization, lamin disassembly and chromosome condensation, are independent processes. *Cell*, **48**, 219–30.

Nicklas, R.B. (1988) The forces that move chromosomes in mitosis. *Ann. Rev. Biophys. Biophys. Chem.*, **17**, 431–49.

Otegui, C. and Patterson, R.J. (1981) RNA metabolism in isolated nuclei: processing and transport of immunoglobulin light chain sequences. *Nucl. Acids Res.*, 9, 4767–81.

Paine, P.L. (1987) The *in vivo* cytomatrix: minimally disturbed systems, in *Modern Cell Biology* (ed. B.H. Satir) Vol. 5, Alan R. Liss, New York, pp. 169–75.

Paine, P.L. (1988) Nuclear protein accumulation: envelope transport or phase affinity mechanisms? *Cell Biol. Int. Rep.*, 12, 691–708.

Paine, P.L. and Feldherr, C.M. (1972) Nucleocytoplasmic exchange of macro-molecules. *Exp. Cell Res.*, 74, 81–98.

Paine, P.L. and Horowitz, S.B. (1980) The movement of material between the nucleus and cytoplasm, in *Cell Biology: A Comprehensive Treatise* (eds D.M. Prescott and L. Goldstein), Academic Press, New York, pp. 299–338.

Paine, P.L., Moore, L.C. and Horowitz, S.B. (1975) Nuclear envelope permeability. *Nature (Lond.)*, 254, 109–14.

Palayoor, T., Schumm, D.E. and Webb, T.E. (1981) Transport of functional messenger RNA from liver nuclei in a reconstituted cell-free system. *Biochim. Biophys. Acta*, 654, 201–10.

Pederson, T. (1983) Nuclear RNA-protein interactions and mRNA processing. *J. Cell Biol.*, 87, 1321–26.

Peters, R. (1986) Fluorescence microphotolysis to measure nucleocytoplasmic transport and intracellular mobility. *Biochim. Biophys. Acta*, 864, 305–59.

Picard, D. and Yamamoto, K.R. (1987) Two signals mediate hormone-dependent nuclear localization of the glucocorticoid receptor. *EMBO J.*, 6, 3333–40.

Prochnow, D., Riedel, N., Agutter, P.S. *et al.* (1990) Poly(A) binding proteins located at the inner surface of resealed nuclear envelopes. *J. Biol. Chem.*, in press.

Purrello, F., Vigneri, R., Clawson, G.A. *et al.* (1982) Insulin stimulation of nucleoside triphosphatase activity in isolated nuclear envelopes. *Science*, 216, 1005–7.

Rao, M.V.D. and Prescott, D.M. (1967) Return of RNA into the nucleus after mitosis. *J. Cell Biol.*, 35, 109a.

Richardson, J.C.W. and Agutter, P.S. (1980) The relationship between the nuclear membranes and the endoplasmic reticulum in interphase cells. *Biochem. Soc. Trans.*, 8, 459–65.

Richardson, W.D., Mills, A.D., Dilworth, S.M. *et al.* (1988) Nuclear protein migration involves two steps: rapid binding at the nuclear envelope followed by slower translocation through nuclear pores. *Cell*, 52, 655–64.

Riedel, N., Fasold, H., Bachmann, M. *et al.* (1987) Permeability measurements in closed vesicles from rat liver nuclear envelopes. *Proc. Natl Acad. Sci. USA*, 84, 3540–4.

Roberts, B.C., Richardson, W.D. and Smith, A.E. (1987) The effect of protein context on nuclear location signal function. *Cell*, 50, 465–75.

Rottmann, M., Schröder, H-C., Gramzow, M. *et al.* (1987) Specific phosphorylation of proteins in pore-complex laminae from the sponge *Geodia cydonium* by the homologous aggregation factor and phorbol ester. Role of protein kinase C in the phosphorylation of DNA topoisomerase II. *EMBO J.*, 6, 3939–44.

Schindler, M. and Jiang, L-W. (1986) Nuclear actin and myosin as control elements in nucleocytoplasmic transport. *J. Cell Biol.*, 102, 859–62.

Schmidt-Zachmann, M.S., Hügle, B., Scheer, U. *et al.* (1987) Identification and localization of a novel nucleolar protein of high molecular weight by a mono-clonal antibody. *Exp. Cell Res.*, 153, 327–46.

Schneider, J.H. (1959) Factors affecting the release of nuclear RNA from the nucleus *in vitro*. *J. Biol. Chem.*, **234**, 2728–32.

Schröder, H-C., Bachmann, M. and Müller, W.E.G. (1989) *Methods for Investigating Nucleo-cytoplasmic Transport of RNA*, Gustav Fisher Verlag, Stuttgart and New York.

Schröder, H-C, Rottmann, M., Bachmann, M. *et al.* (1986a) Purification and characterization of the major NTPase from rat liver nuclear envelopes. *J. Biol. Chem.*, **261**, 663–8.

Schröder, H-C., Rottmann, M., Bachmann, M. *et al.* (1986b) Proteins from rat liver cytosol which stimulate mRNA transport: purification and interactions with the nuclear envelope mRNA translocation system. *Eur. J. Biochem.*, **159**, 51–9.

Schröder, H-C., Rottmann, M., Wenger, R. *et al.* (1988a) Studies on protein kinases involved in regulation of nucleocytoplasmic mRNA transport. *Biochem. J.*, **252**, 777–90.

Schröder, H-C., Trölltsch, D., Friese, U. *et al.* (1987) Mature messenger RNA is selectively released from the nuclear matrix by an ATP/deoxy-ATP-dependent mechanism sensitive to topoisomerase inhibitors. *J. Biol. Chem.*, **262**, 8917–25.

Schröder, H-C., Diehl-Seifert, B., Rottmann, M. *et al.* (1988b) Functional dissection of nuclear envelope mRNA translocation system: effects of phorbol ester and a monoclonal antibody recognizing cytoskeletal structures. *Arch. Biochem. Biophys.*, **261**, 394–404.

Schulz, B. and Peters, R. (1987) Nucleocytoplasmic protein traffic in single mammalian cells studied by fluorescence microphotolysis. *Biochim. Biophys. Acta*, **930**, 419–31.

Schumm, D.E., Hanausek-Walasek, M., Yannarell, A. *et al.* (1977) Changes in nuclear RNA transport incident to carcinogenesis. *Eur. J. Cancer*, **13**, 139–47.

Schumm, D.E., Niemann, M.A., Palayoor, T. *et al.* (1979) *In vivo* equivalence of a cell-free system from rat liver for ribosomal RNA processing and transport. *J. Biol. Chem.*, **254**, 12126–30.

Schweiger, A. and Kostka, G. (1984) Concentration of particular high molecular mass phosphoprotein in rat liver nuclei and nuclear matrix decreases following inhibition of RNA synthesis with α-amanitin. *Biochim. Biophys. Acta*, **782**, 262–8.

Sealy, L., Cotten, M. and Chalkley, R. (1986) *Xenopus* nucleoplasmin: egg vs oocyte. *Biochemistry*, **25**, 3064–72.

Senior, A. and Gerace, L. (1988) Integral membrane proteins specific to the inner nuclear membrane and associated with the nuclear lamina. *J. Cell Biol.*, **107**, 2029–36.

Sevaljevic, L., Brajanovic, N. and Trajkovic, D. (1982) Cortisol induced stimulation of nuclear matrix protein phosphorylation. *Mol. Biol. Rep.*, **8**, 225–32.

Shearer, R.W. (1974) Specificity of chemical modification of RNA transport by liver carcinogens in the rat. *Biochemistry*, **13**, 1764–9.

Skoglund, U., Andersson, J., Strandberg, B. *et al.* (1983) Three-dimensional structure of a specific pre-messenger RNP particle established by electron microscope tomography. *Nature*, **319**, 560–4.

Smith, C.D. and Wells, W.W. (1984) Solubilization and reconstitution of a nuclear envelope associated ATPase: synergistic activation by RNA and polyphosphoinositides. *J. Biol. Chem.*, **259**, 11890–4.

Smith, D.E., Gruenbaum, Y., Berrios, M. *et al.* (1987) Biosynthesis and interconversion of *Drosophila* nuclear lamin isoforms during normal growth and in response to heat shock. *J. Cell Biol.*, **105**, 771–80

Smith, H.C., Ochs, R.L., Fernandez, E.A. *et al.* (1986) Macromolecular domains containing nuclear P-107 and UsnRNP protein P-28: further evidence for an *in situ* nuclear matrix. *Mol. Cell Biochem.*, **70**, 151–68.

Snow, C.M., Senior, A. and Gerace, L. (1987) Monoclonal antibodies identify a group of nuclear pore complex glycoproteins. *J. Cell Biol.*, **104**, 1143–56.

Steitz, J.A. (1988) 'Snurps', *Sci. Am.*, **258**, no. 6, 36–41.

Stevens, B.J. and Swift, H. (1966) RNA transport from nucleus to cytoplasm in *Chironomus* salivary glands. *J. Cell Biol.*, **31**, 55–7.

Sugawa, H. and Uchida, T. (1985) Inhibition of RNA nucleocytoplasmic translocation by antinucleus antibody. *Biochem. Biophys. Res. Commun.*, **127**, 864–70.

Swanson, J.A. and McNeil, P.L. (1987) Nuclear reassembly excludes large molecules. *Science*, **238**, 548–50.

Tobian, J.A., Castano, J.G. and Zasloff, M.A. (1984) RNA nuclear transport in *Xenopus laevis* oocytes: studies with human initiator tRNA-met point mutants. *J. Cell Biol.*, **99**, 232a.

Tobian, J.A., Drinkard, L. and Zasloff, M.A. (1985) tRNA nuclear transport: defining the critical regions of the human initiator tRNA-met by point mutagenesis. *Cell*, **43**, 415–22.

Turner, W.A., Taylor, J.D. and Tchen, T.T. (1979) Melanocyte stimulating hormone stimulation of nuclear envelope blebbing. *Pigment Cell*, **4**, 50–5.

Unwin, P.N.T. and Milligan, R.A. (1982) A large particle associated with the nuclear pore complex. *J. Cell Biol.*, **93**, 63–75.

Vale, R.D. (1987) Intracellular transport using microtubule-based motors. *Annu. Rev. Cell Biol.*, **3**, 347–78.

Van Eekelen, C.A.G. and van Venrooij, W.J. (1981) HnRNA and its attachment to a nuclear matrix. *J. Cell Biol.*, **88**, 554–63.

Van Eekelen, C.A.G., Rieman, T. and van Venrooij, W.J. (1981) Specificity of the interaction of HnRNA and mRNA with proteins as revealed by *in vivo* cross-linking. *FEBS Lett.*, **130**, 223–6.

Van Venrooij, W.J., Sillekens, P.T.G., van Eekelen, C.A.G. *et al.* (1981) On the association of mRNA with the cytoskeleton in uninfected and adenovirus infected human KB cells. *Exp. Cell Res.*, **135**, 79–92.

Verheijen, R., Kuijpers, H., Vooijs, P. *et al.* (1986) Distribution of the 70K U1 RNA-associated protein during interphase and mitosis: correlation with other U RNP particles and proteins of the nuclear matrix. *J. Cell Sci.*, **86**, 173–90.

Virtanen, I. (1977) Identification of concanavalin A binding glycoproteins in rat liver cell nuclear membranes. *Biochem. Biophys. Res. Commun.*, **78**, 1411–17.

Vorbrodt, A. and Maul, G.G. (1980) Cytochemical studies on the relation of nucleoside triphosphatase activity to ribonucleoproteins in isolated rat liver nuclei. *J. Histochem. Cytochem.*, **28**, 27–35.

Wolosewick, J.J. and Porter, K.R. (1976) Microtrabecular lattice of the cytoplasmic ground substance. *J. Cell. Biol.*, **82**, 114–39.

Worman, H.J., Yuan, J., Blobel, G. *et al.* (1988) A lamin B receptor in the nuclear envelope. *Proc. Natl Acad. Sci. USA*, **85**, 8531–4.

Wozniak, R.W., Bartnik, E. and Blobel, G. (1989) Primary structure analysis of an integral membrane glycoprotein of the nuclear pore. *J. Cell Biol.*, **108**, 2083–92.

Wunderlich, F., Giese, G. and Herlan, G. (1984) Thermal down-regulation of exportable rRNA in nuclei. *J. Cell Physiol.*, **120**, 211–18.

Yamaizumi, M., Uchida, T., Okada, Y. *et al.* (1978) Rapid transfer of non-histone chromosomal proteins to the nucleus of living cells. *Nature (Lond.)*, **273**, 782–4.

Yamasaki, L., Kanda, P. and Lanford, R.E. (1989) Identification of four nuclear transport signal-binding proteins that interact with diverse transport signals. *Mol. Cell Biol.*, **9**, 3028–36.

Yoneda, Y., Imamoto-Sonobe, N., Matsuoka, Y. *et al.* (1988) Antibodies to asp-asp-glu-asp can inhibit transport of nuclear proteins into the nucleus. *Science*, **242**, 275–8.

Zaboikin, M.M., Lichtenstein, A.V. and Shapot, V.S. (1987) Interrelationships of structural, metabolic and adhesive properties of nuclear and cytoplasmic RNAs. *Biokhimiiia*, **52**, 794–805.

Zasloff, M. (1983) tRNA transport from the nucleus in a eucaryotic cell: carrier-mediated process. *Proc. Natl Acad. Sci. USA*, **80**, 6436–40.

Index